MECHANICAL ENGINEERING THEORY AND APPLICATIONS

COMPUTATIONAL FLUID DYNAMICS

ADVANCES IN RESEARCH AND APPLICATIONS

MECHANICAL ENGINEERING THEORY AND APPLICATIONS

Additional books and e-books in this series can be found on Nova's website under the Series tab.

MECHANICAL ENGINEERING THEORY AND APPLICATIONS

COMPUTATIONAL FLUID DYNAMICS

ADVANCES IN RESEARCH AND APPLICATIONS

JAMES S. HUTCHINSON
EDITOR

Copyright © 2021 by Nova Science Publishers, Inc.

All rights reserved. No part of this book may be reproduced, stored in a retrieval system or transmitted in any form or by any means: electronic, electrostatic, magnetic, tape, mechanical photocopying, recording or otherwise without the written permission of the Publisher.

We have partnered with Copyright Clearance Center to make it easy for you to obtain permissions to reuse content from this publication. Simply navigate to this publication's page on Nova's website and locate the "Get Permission" button below the title description. This button is linked directly to the title's permission page on copyright.com. Alternatively, you can visit copyright.com and search by title, ISBN, or ISSN.

For further questions about using the service on copyright.com, please contact:
Copyright Clearance Center
Phone: +1-(978) 750-8400   Fax: +1-(978) 750-4470   E-mail: info@copyright.com.

NOTICE TO THE READER

The Publisher has taken reasonable care in the preparation of this book, but makes no expressed or implied warranty of any kind and assumes no responsibility for any errors or omissions. No liability is assumed for incidental or consequential damages in connection with or arising out of information contained in this book. The Publisher shall not be liable for any special, consequential, or exemplary damages resulting, in whole or in part, from the readers' use of, or reliance upon, this material. Any parts of this book based on government reports are so indicated and copyright is claimed for those parts to the extent applicable to compilations of such works.

Independent verification should be sought for any data, advice or recommendations contained in this book. In addition, no responsibility is assumed by the Publisher for any injury and/or damage to persons or property arising from any methods, products, instructions, ideas or otherwise contained in this publication.

This publication is designed to provide accurate and authoritative information with regard to the subject matter covered herein. It is sold with the clear understanding that the Publisher is not engaged in rendering legal or any other professional services. If legal or any other expert assistance is required, the services of a competent person should be sought. FROM A DECLARATION OF PARTICIPANTS JOINTLY ADOPTED BY A COMMITTEE OF THE AMERICAN BAR ASSOCIATION AND A COMMITTEE OF PUBLISHERS.

Additional color graphics may be available in the e-book version of this book.

Library of Congress Cataloging-in-Publication Data

ISBN: 978-1-53619-756-3

Published by Nova Science Publishers, Inc. † New York

CONTENTS

Preface		vii
Chapter 1	Phase Change Simulations in Particle Flow Using Computational Fluid Dynamics *Laurie A. Florio*	1
Chapter 2	CFD-Based Design of Novel Green Flash Ironmaking Reactors *H. Y. Sohn*	55
Chapter 3	Modeling and Computational Fluid Dynamics (CFD) Simulation of CO_2 Absorption Using Mono-Ethanol Amine (MEA) Solution in a Hollow Fiber Membrane (HFM) Contactor *Ehsan Kianfar and Sajjad Golchin Khazari*	97
Chapter 4	Simulation of High-Temperature Air Effects in Hypersonic Flows *Yu. Dobrov, A. Karpenko and K. Volkov*	125
Index		171

PREFACE

This monograph consists of four chapters, each of which present new research in the field of computational fluid dynamics (CFD). Chapter One describes volume of fluid and moving mesh approaches to modeling phase change in a system with particles or droplets in a CFD environment. Chapter Two describes a novel CFD-based design of potential industrial reactors for flash ironmaking. Chapter Three presents a study wherein CFD was used to simulate a hollow fiber membrane contactor for the absorption of carbon dioxide from the air by mono-ethanol amine. Chapter Four describes the use of graphical processor units (GPUs) for the simulation of high-speed and high-temperature flows in CFD.

Chapter 1 - Phase change is critical to many of the phenomena related to the operation and control of various engineering systems. Phase change may be used to limit temperature levels, may be implemented to alter the concentrations of or filter system components, may lead to conditions that cause erosion and fouling, and may serve as the foundation of many manufacturing, industrial, and natural processes. An understanding of the relationships between the key system parameters and the phase change that develops is crucial to improving the understanding of phase change and therefore improving the ability to harness the related phenomena for useful purposes. Computational fluid dynamics based modeling techniques hold certain advantages over experimental testing since local temperatures, pressures, and velocities can be readily determined and traced over time and visualization of the flow conditions can be obtained. Hence, the state of the phase change process and the resulting local changes in the system geometry, flow and temperature fields, and distribution of the gas species can be captured in a level of detail that is difficult to achieve with physical testing. In this work, volume of fluid and/or moving mesh approaches to modeling phase change in a system with particles/droplets in a computational fluid dynamics environment are described. Methods are developed and applied to simulate ablative phase change from particle surfaces, evaporation of liquid droplets, and the solidification of metal on the surface of cold solid metal particles as representative of the modeling capability advancement. The utility of the methods is demonstrated and areas for further work and application are discussed.

Chapter 2 - A critical problem facing the steel industry is to develop a new technology to produce iron with significantly reduced energy consumption and greenhouse gas emissions. The development of a novel ironmaking technology based on direct utilization of fine iron ore concentrate in a flash reactor is summarized. The CFD-based design of potential industrial reactors for flash ironmaking is described. Overall, this work has shown that the size of the reactor used in the novel Flash Ironmaking Technology (FIT) can be quite reasonable vis-à-vis the blast furnaces. As an example, a flash reactor of 12 m diameter and 35 m with a single burner operating at atmospheric pressure would produce 1.0 million tons of iron per year. The height can be further reduced by using multiple burners and/or preheating the feed gas. The CFD-based design of potential industrial reactors for flash ironmaking pointed to a number of features that an industrial reactor should incorporate. The flow field should be designed in such a way that a larger portion of the reactor is used for the reduction reaction but at the same time excessive collision of particles with the wall must be avoided. Further, a large diameter-to-height ratio should be used from the viewpoint of decreased heat loss. This may require the incorporation of multiple burners and solid feeding ports on the reactor roof.

Chapter 3 - In this study, computational fluid dynamics technique (CFD) was used to simulate a hollow fiber membrane (HFM) contactor for the absorption of carbon dioxide from the air by mono-ethanol amine (MEA). The governing equations of this process include the equations of continuity, momentum and mass transfer in areas related to the shell membrane and tube and by taking into account the relevant boundary conditions and assumptions and using the software COMSOL, the equations were solved. By examining and comparing the results with experimental data presented in the literature, the accuracy of simulation results was confirmed. The influences of important parameters in the process including wetness of membrane, the volumetric flow rate of gas, liquid volumetric flow rate and temperature effects were studied. Results show that with increasing the volumetric flow rate of the liquid absorber, the mass transfer rate of carbon dioxide to absorption increases because of the carbon dioxide concentration gradient in the gas phase and liquid increases. Because of the reduced residence time in the membrane contactor, with increasing the volumetric flow rate of gas the amount of removed CO_2 in contact reduces so that with increasing the gas flow rate from 0.6 to 1.4, removed CO_2 reduces only 18 percent. The other factors are the solvent temperature and wetness. With increasing the solvent temperature, carbon dioxide removal efficiency is significantly increased. Wettability has a strong negative effect on the efficiency of the membrane contactor. Thus with increasing wettability, the efficiency of the membrane suddenly finds a sharp decrease. In some cases, increased wettability efficiency is close to zero.

Chapter 4 - Development and implementation of methods and tools that adequately model fundamental physics and allow credible physics-based optimization for future operational hypersonic vehicle systems are becoming more important due to requirements of ensuring their flight safety. The methods of computational fluid dynamics (CFD) are extensively applied in design and optimization of hypersonic vehicles to get more insight into complex flowfields. Computer simulation is particularly attractive due to its relatively low cost and its ability to deliver data that cannot be measured or observed. Flow discontinuities, high gradients of flow quantities, turbulence effects, flow separation and other flow features impose great demands on the underlying numerical methods. The use of Graphics Processor Units (GPUs) is a cost effective way of improving substantially the performance in CFD applications. GPU platforms make it possible to achieve speedups of an order of magnitude over a standard CPU in many CFD applications. The parallel capabilities of in-house compressible CFD code for hypersonic flow simulations are assessed and successful design of a highly parallel computation system based on GPUs is demonstrated. Possibilities of the use of GPUs for the simulation of high-speed and high-temperature flows are discussed. The results obtained are generally in a reasonable agreement with the available experimental and computational data, although some important sensitivities are identified.

In: Computational Fluid Dynamics
Editor: James S. Hutchinson

ISBN: 978-1-53619-756-3
© 2021 Nova Science Publishers, Inc.

Chapter 1

PHASE CHANGE SIMULATIONS IN PARTICLE FLOW USING COMPUTATIONAL FLUID DYNAMICS

Laurie A. Florio[*]
Armaments Technology and Evaluation Division,
US ARMY Armament Graduate School,
US ARMY DEVCOM AC, Picatinny Arsenal, NJ, US

ABSTRACT

Phase change is critical to many of the phenomena related to the operation and control of various engineering systems. Phase change may be used to limit temperature levels, may be implemented to alter the concentrations of or filter system components, may lead to conditions that cause erosion and fouling, and may serve as the foundation of many manufacturing, industrial, and natural processes. An understanding of the relationships between the key system parameters and the phase change that develops is crucial to improving the understanding of phase change and therefore improving the ability to harness the related phenomena for useful

---

[*] Corresponding Author's Email: laurie.a.florio.civ@mail.mil.

purposes. Computational fluid dynamics based modeling techniques hold certain advantages over experimental testing since local temperatures, pressures, and velocities can be readily determined and traced over time and visualization of the flow conditions can be obtained. Hence, the state of the phase change process and the resulting local changes in the system geometry, flow and temperature fields, and distribution of the gas species can be captured in a level of detail that is difficult to achieve with physical testing. In this work, volume of fluid and/or moving mesh approaches to modeling phase change in a system with particles/droplets in a computational fluid dynamics environment are described. Methods are developed and applied to simulate ablative phase change from particle surfaces, evaporation of liquid droplets, and the solidification of metal on the surface of cold solid metal particles as representative of the modeling capability advancement. The utility of the methods is demonstrated and areas for further work and application are discussed.

Keywords: phase change, evaporation, condensation, ablation, solidification, melting, computational fluid dynamics, particles, droplets, compressible flow

INTRODUCTION

Phase change often plays a prominent role in controlling thermal conditions in a system (Faghri & Zhang, 2020), (Sun, 2014), (Ray, Durst, & Kumar, 2007), in filtering or controlling concentrations of fluid components, in the development of fouling or erosion, and in manufacturing, industrial, and natural processes. The latent heat required to transition from solid to gas, liquid to gas, or solid to liquid utilizes system energy for this phase change process, delaying further temperature increases. In a similar manner, phase changes and related chemistry at a surface of a solid or droplet transfer mass, momentum, and energy and affect the system conditions. Evaporation or condensation may serve to remove certain fluids or particulates/drops carried with the flow from a flow stream or alter fluid component configurations. Melting may result in material removal or erosion of surfaces and fouling could develop as solidification then occurs along a flow surface.

Phase change involves a set of interdependent phenomena and effects. Surfaces or interfaces where the change develops may regress or may grow with the interfaces varying in phase-type, geometry, or composition with time. As the phases and composition changes, the thermodynamic and transport properties of the fluids and solids vary, potentially affecting the flow field and the thermal conditions. The thermal and mechanical interactions between the phases can also affect the phase change processes or may be altered by these processes. The phase change or chemical effects are driven by the local thermal, composition, and flow effects. While physical experiments are necessary for determining parameters and confirming behaviors, computational models offer a means for detailed investigations, under controlled conditions, into the phase change phenomena of interest and the relationships between the system parameters and the effects of the phase change (Ray, Durst, & Kumar, 2007). An improved understanding of the underlying phenomena can be gained from examination and analysis of the computational fluid dynamics model results. This understanding can then be applied to develop more effective means of controlling the phase change phenomena for an intended purpose.

This chapter presents methods for implementing computational fluid dynamics techniques to simulate three major types of phase change related phenomena: ablative phase change, evaporative/condensation phase change, and melting/solidification phase change. For each of these cases, the phase change studied initiates at the surface of a particle or droplet interface. After background information for each of these three phase change types is presented, the specific computational fluid dynamics methods are described and the detailed information that can be obtained from the output of the simulations using these methods are presented. Finally, the need for further development of the phase change computational fluid dynamics based methods, particularly, related to particle flow are described.

PHASE CHANGE MODELING BACKGROUND

Means to simulate phase change has been an area of active research and continual development. As computational tools, methods, and capabilities have grown, the ability to model phase change conditions in a fluid has expanded. A review of the common methods for simulating ablation, evaporation/condensation, melting/solidification is provided below as a basis for the modeling methods associated with the particles and droplets that are the focus of this work.

Ablative Related Phase Change

The ablative phase change is commonly utilized to protect a surface from the effects of high temperature with the ablative material placed on the surface that would otherwise be directly exposed to the high heating rates or temperatures. Applications include the exterior surfaces of systems moving at high speed such as re-entry vehicles or interior surfaces subject to high temperature flows such as rocket motor surfaces. (See Figure 1).

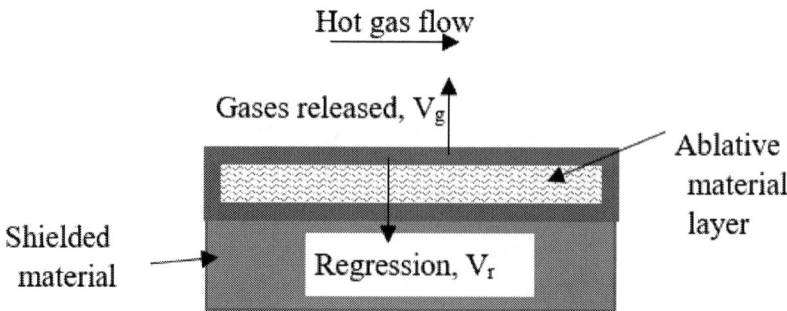

Figure 1. Ablative phase change.

The ablative materials are typically of low thermal conductivity and high melt temperature, commonly carbon and/or glass, held within a polymer binder. When the sacrificial ablative material is exposed to a high heat flux, the material may begin to melt with the melt carried away with the flow stream or vaporized due to the high temperatures or conditions may be such that the ablative material sublimates. Because the ablative process is ongoing once the material reaches the phase change temperature and continues be heated, no special controls to start or stop the thermal control effects are required. The ablative material, however, is consumed in the process, and therefore has a finite useful life (Faghri & Zhang, 2020) (Nellis & Klein, Heat Transfer, 2012).

A number of different methods to model the ablative processes have been developed and applied. The work of Park and Ahn (Park & Ahn, 1999) and Zhluktov-Abe (Zhluktov & Abe, 1999) are among the more commonly applied models for carbon based ablative materials. These and other models have then been applied in different manners to study a variety of ablative applications. Chen and Milos (Chen & Milos, 2004) formulated specific ablation boundary conditions to be used in numerical models of carbon based ablation and applied these models to conditions of heated air flow moving over a carbon based surface. Results with the use of the Park-Ahn and Zhuluktov-Abe models are compared. Johnston, Gnoffo, and Mazaheri (Johnston, Gnoffo, & Mazaheri, 2012) report on a method to couple the surface reactions associated with ablation directly to models for the gas flow. Alanyalioglu (Alanyalioglu, 2019) has worked to develop modeling methods for ablation effects associated with the surface of a rocket motor using the more general commercial code FLUENT©. Mei et al. have combined surface models and gas chemistry to simulate the conditions for hypersonic reentry flow around a thin spherical cap and incorporate the surface regression effects using ABAQUS© (Mei, Chenging, Fan, & Wang, 2020). The effects of using different surface and gas chemistry models are explored in this work for conditions at this external surface.

In the current work, the focus is on carbon ablative material and so the previous work related to carbon ablative material is discussed here. The species and rate of gas generation are based on the work of Park and Ahn (Park & Ahn, 1999). The oxidation and nitridation reactions considered are:

$$O + C_{(s)} \rightarrow CO \tag{1}$$

$$O_2 + 2C_{(s)} \rightarrow 2CO \tag{2}$$

$$N + C_{(s)} \rightarrow NO \tag{3}$$

Based on these reactions, where here the C_3 produced from the sublimation of carbon is neglected, Chen and Milos (Chen & Milos, 2004) formulated expressions for the mass flux loss rate of the carbon, $\dot{m}_c$, in kg/m²-s, as given in Eq.(4) where each term on the right of the equations results from the interactions of the carbon with the gas species O, O_2, and N, in $\dot{m}_1, \dot{m}_2,$ and $\dot{m}_3$ respectively.

$$\dot{m}_c = \dot{m}_1 + \dot{m}_2 + \dot{m}_3 \tag{4}$$

The carbon gas mass flux rates in kg/m²-s associated with the three gas species are given in Eqs. 5 through 7 where Y_i is the mass fraction of species i, ρ_{gas} is the density of the gas, M_i is the molecular weight of species i, v_i are velocities defined below and β_i are parameters defined in Eq. (8). For the O:

$$\dot{m}_1 = \rho_{gas} Y_o v_o \beta_o \frac{M_C}{M_o} \tag{5}$$

For the O_2:

$$\dot{m}_2 = 2\rho_{gas} Y_{O_2} v_{O_2} \beta_{O_2} \frac{M_C}{M_{O_2}} \tag{6}$$

For the N:

$$\dot{m}_3 = 2\rho_{gas} Y_N v_N \beta_N \frac{M_C}{M_N} \tag{7}$$

The v_i expressions are velocities defined in Eq 8 where k is the Boltzmann constant, 1.38×10^{-23} J/(K-molecule), T_w is the wall temperature in K, and m_i is the mass of a single molecule of gas species i in kg (Kuo, 2005).

$$v_i = \sqrt{\frac{kT_w}{2\pi m_i}} \qquad (8)$$

The single molecule mass of species i can be found from:

$$m_i = M_i / N_A \qquad (9)$$

with $N_A=6.0225 \times 10^{26}$ molecules/kg-mole and M_i, the molecular weight of species i in kg/kg-mol. The values of β_i from Chen and Milos are given below for each of the gas species in Eqs. 5 through 7. For O:

$$\beta_O = 0.63 exp\left(-\frac{1160}{T_W}\right) \qquad (10)$$

For O_2:

$$\beta_{O_2} = 0.5 \qquad (11)$$

For N:

$$\beta_N = 0.3 \qquad (12)$$

Now, with the mass of carbon "blowing" rates specified, the regression rate of the carbon ablative surface can be calculated along with the net "blowing" velocity of the gases produced from the gas-ablative surface interactions. The regression rate of the ablative surface, V_r, in the direction normal to the ablative surface as shown in Figure 1 is given by:

$$V_r = \frac{\dot{m}_c}{\rho_c} \qquad (13)$$

where ρ_C is the density of the ablative material in kg/m³. From conservation of mass, then, the mean velocity of the gas products entering the gas flow side of the system, V_g, is given in Equation 14 (Figure 1) where ρ_g is the density of the gas mixture entering at the wall temperature and pressure conditions:

$$V_g = \frac{\rho_c V_r}{\rho_g} \tag{14}$$

The mass flux generation rates of the gas species involved can also be calculated. The mass flux rate of CO (kg/m²-s), is given by:

$$\dot{m}_{CO} = \dot{m}_1 \frac{M_{CO}}{M_C} + \dot{m}_2 \frac{M_{CO}}{M_C} \tag{15}$$

The mass flux rate of O (kg/m²-s), is given by:

$$\dot{m}_O = -\dot{m}_1 \frac{M_{CO}}{M_C} \tag{16}$$

The mass flux rate of O_2 (kg/m²-s), is given by:

$$\dot{m}_{O_2} = -\dot{m}_2 \frac{M_{O_2}}{2M_C} \tag{17}$$

The mass flux rate of N (kg/m²-s), is given by:

$$\dot{m}_N = -\dot{m}_3 \frac{M_N}{M_C} \tag{18}$$

These expressions come from Chen and Milos.

Once this "pyrolysis" gas injection resulting from the solid carbon surface – gas interactions is introduced into the main gas flow, the gas species may react with the gas species already moving over the ablative material surface. In this work, a reduced chemical kinetic mechanism based on the work of Gupta and et al. (Gupta, Yos, Thompson, & Lee, 1990) is utilized to model the chemical reactions that may be taking place in the gas mixture. The gas species considered in the model are CO, CO_2, O_2, O, N, N_2, and NO. The forward and backward rates for a seven reaction mechanism are provided in Table 1 along with the third body efficiencies for those that are not equal to 1.

Table 1. Gas species reaction in formation from Gupta et al.

Reaction		Forward Rate $k_f = AT^n \exp(-E_a/(R_uT))(m^3/\text{kg-mole-s})$			Backward Rate $k_b = AT^n \exp(-E_a/(R_uT))(m^3/\text{kg-mole-s})$ or $(m^6/(\text{kg-mole})^2\text{-s})$			Third Body Efficiencies ≠1
		A	n	E_a	A	n	E_a	
1	$O_2 \rightarrow O+O$	3.61E+15	-1.00	4.94E+08	3.01E+09	-0.50	0.0000E+00	O=25; O_2=9; N_2=2
2	$N_2 \rightarrow N+N$	1.92E+14	-0.50	9.40E+08	1.09E+10	-0.50	0.0000E+00	N=0; N_2=2.5
3	$N_2+N \rightarrow N+N+N$	4.15E+19	-1.50	9.43E+08	2.32E+15	-1.50	0.0000E+00	N/A
4	$NO \rightarrow N+O$	3.97E+17	-1.50	6.26E+08	1.01E+14	-1.50	0.0000E+00	N=20; NO=20; O_2=20
5	$NO+O \rightarrow N+O_2$	3.18E+06	0.00	1.64E+08	9.63E+08	0.50	2.9930E+07	N/A
6	$N_2+O \rightarrow NO+N$	6.75E+10	0.00	3.12E+08	1.50E+10	0.00	0.0000E+00	N/A
7	$CO+O_2 \rightarrow CO_2+O$	1.00E+09	0.00	1.99E+08	1.50E+10	0.00	0.0000E+00	N/A

The modeling of the surface ablation requires the determination of the gas species production rate, the type of gases generated as a result of the ablation, the determination of the appropriate mass, species, energy and momentum introduced into the main gas flow stream moving over the ablative surface, the regression of the ablative material surface, and the chemical reactions occurring in the gases moving over the heated surface. A method/technique to incorporate and model these phenomena in a general CFD model is introduced later in this chapter.

Evaporation/Condensation Phase Change

The second phase change related phenomena studied in this chapter is evaporation/condensation type phase change. Like ablation, evaporation can also be used to control the thermal conditions of a system through the cooling of the working fluid, the cooling of the surfaces over which the evaporating fluid is moving or a surface upon which the fluid is applied, or can be used to increase the concentration of a given component in a liquid mixture by removing a certain liquid substance or by removing the liquid from a liquid-particle mixture (Faghri & Zhang, 2020). Evaporation and vaporization also play a role in the combustion of some substances as well (Law, 2006) (Kuo, 2005). The evaporative phenomena initiates at the interface of a liquid as the temperature at the interface approaches the saturation temperature at the given pressure. The heating may be due to contact with a heated solid wall or may be due to interactions with a hot gas/fluid surrounding the droplet (Faghri & Zhang, 2020). The energy required to cause the phase change from liquid to gas is the latent heat of vaporization. The droplet surface temperature is maintained at the saturation temperature until the local phase change has been completed. With the ablation process, the liquid that may form as a result of the phase change is either promptly vaporized/sublimates or is frequently promptly carried away by the external gas flow. With evaporation, and also during condensation, however, the heat conduction and diffusion effects within the droplet may cause the development of convection currents within the droplet, affecting the temperature distribution, fluid properties, material distribution, and phase change.

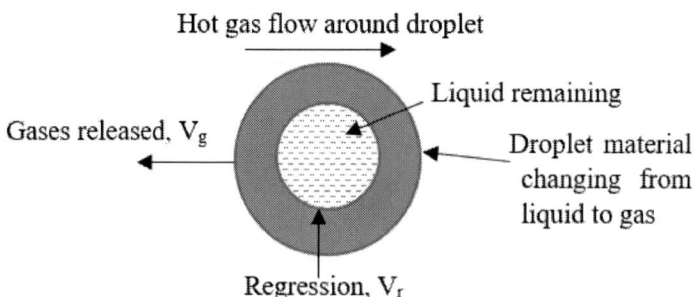

Figure 2. Droplet evaporation.

Just like evaporation processes, condensation processes develop at interfaces and can affect the thermal conditions, can be required as part of a thermal control cycle involving evaporation and condensation, or can be used to alter the concentration of particular substances within a fluid stream. When a saturated vapor, in a gas state, contacts a cooler surface, particle, or a cooler fluid stream, phase change from the vapor to the liquid can develop. In such cases, heat transfer develops from the vapor to the wall/fluid as a result of the phase change with the heat transferred from the droplet equal to the heat of vaporization, with the heat flow opposite to that for the evaporation discussed earlier (Faghri & Zhang, 2020).

The applications for modeling evaporation and condensation span a range of fields. The few studies mentioned below illustrate various applications of the evaporation/condensation change phenomena. Khakpour and Sayed-Yagoobi (Khakpour & Seyed-Yagoobi, 2015) have modeled the evaporation occurring in a liquid film, carrying micro-encapsulated phase change materials, moving over a flat plate. Qui et al. (Qui, Cai, Wu, Yao, & Jiang, 2015) report on the results of models of the condensation occurring in curved tubes. Zhang et al. use a CFD method based on the diffusion effects of the phases to model the evaporation and vaporization of a droplet in lower speed gas flow (Zhang, Zhang, & Zhang, 2012). In all of these studies, the detailed information regarding the droplet shapes, deformation, rates of vaporization/evaporation, and the changes in the heat transfer or flow that result can be readily tabulated through the numerical CFD simulations, providing information that would be difficult to obtain through standard testing methods.

Evaporation modeling will be the focus of this work, though similar methods, with the opposite heat flow direction, can be applied to modeling condensation effects. Analytical models for simplified systems such as a spherical droplet surrounded by quiescent hot gas are presented in Faghri and Zhang (Faghri & Zhang, 2020), Kuo (Kuo, 2005), and Law (Law, 2006). For CFD based methods, the evaporation type models are typically based on the use of a volume of fluid (VOF) formulation where the volume fraction of the phases is traced over time. Such a formulation allows for the tracing of the droplet interface, droplet deformation, and surface tension effects. For a system containing only the liquid and gas/vapor phases of a given substance, the volume fraction of the liquid, α_l, the volume fraction of the vapor, α_v, must sum to one in each of the computational cells.

$$\alpha_l + \alpha_v = 1 \tag{19}$$

The properties of the fluid in a given computational cell are weighted by these volume fractions, so that property, ϕ, is given by:

$$\phi = \alpha_l \phi_l + \alpha_v \phi_v \tag{20}$$

However, only one temperature, pressure, and velocity, or turbulence parameter is defined for each of the computational cells, i.e., the phases within the computational domain cannot be assigned different velocities for instance. The effects of the surface tension can be included in the simulations with the method developed by Brackbrill et al. (Brackbrill, Kothe, & Zemach, 1992) among the methods that can be used.

In both the surface tension force calculations and in the phase change modeling, the interface between different phases encompasses the cells where the volume fraction of one phase or the other is not equal to one. The interface area, or surface normal, **n**, can be estimated by the gradient in the volume fraction as:

$$\boldsymbol{n} = \nabla \alpha_l \tag{21}$$

This particular normal or interface surface area, based on the liquid volume fraction, is used in some of the formulations for the phase change mass transfer rates.

Using this VOF based approach, many methods have been devised for modeling the evaporation/condensation phenomena. One of the earlier methods is that of Lee (Lee, 1980) where when the cell temperature is greater than the fluid saturation temperature at a given pressure, T_S, a source term is calculated when the liquid volume fraction is non-zero, indicating there is liquid remaining in the computational cell and evaporation should be taking place. The volumetric mass generation of the liquid, $\dot{m}_l$, and the negative of the volumetric mass generation rate of the vapor phase, $\dot{m}_v$, under such conditions are:

$$\dot{m}_l = -\dot{m}_v = -r \alpha_l \rho_l \left(\frac{T - T_S}{T_S}\right); T \geq T_S \tag{22}$$

where T is the computational cell temperature, T_s is the saturation temperature, and the subscript *l* stands for liquid with the subscript v for vapor. When the temperature in the cell is less than the saturation temperature at a given pressure, then, if there is vapor within the computational cell and condensation should be taking place, the volumetric mass generation rate of the vapor phase and the negative of the volumetric mass generation rate of the liquid phase can be found below.

$$\dot{m}_v = -\dot{m}_l = -r\alpha_v\rho_v\left(\frac{T-T_S}{T_S}\right); T < T_S \qquad (23)$$

In each of these cases, a factor r is present. This factor r is to be adjusted to keep the interface temperature near the saturation temperature and is not a physically meaningful parameter. A corresponding energy source must also be prescribed, Keeping the appropriate sign of the vapor volumetric mass generation rate, the energy source, with h_{fg} as the latent heat of vaporization, is:

$$\dot{Q} = \dot{m}_v h_{fg} \qquad (24)$$

A variation in the calculation of the mass transfer between the phases was developed by Welch and Wilson (Welch & Wilson, 2000), among others. Instead of the rates given in Equations 22 and 23, the following expression is used for the volumetric generation of vapor (k_m is the mixture thermal conductivity calculated as in Equation 20):

$$\dot{m}_v = \left(\frac{-k_m \nabla T \cdot \frac{\nabla \alpha_l}{|\nabla \alpha_l|}}{h_{fg}}\right) \qquad (25)$$

Sun (Sun, 2014) reports on a modified method where the net conductive heat flux through the computational cell boundaries in the direction of the interface normal divided by the latent heat is used to formulate the mass transfer. Variations on these basic methods have been developed to try to capture the phase change phenomena with limited use of non-physical parameters.

After calculating the required mass transfer and accompanying sources, additional techniques may be required to model the phase change. For instance, the large differences in the properties of the phases can lead to numerical issues (Shyy, 1994). Thermal/transport properties, the thermodynamic properties, as well as the density of the fluids involved in the phase change and thus the volume occupied by the same mass can be significantly different. To circumvent potential issues, in some of these models, the density of the different phases is held constant in many studies so that the volume occupied by the same mass of material does not change. However, this fixed density level may not be a feasible assumption for many analyses and materials. Volume changes or other methods may be required.

Melting/Solidification Phase Change

The last phase change related phenomena studied in this chapter is the melting and solidification type of phase change. Just as with the evaporation and condensation, the melting and solidification phase changes have a wide variety of applications. Melting and solidification is important in many natural processes, manufacturing processes including additive manufacturing and the effects of frictional heating, and is used for energy storage and thermal control. Solidification on surfaces can lead to fouling and melting can lead to surface erosion, both causing potentially detrimental effects on a system operation. When the temperature of a solid material reaches the melt temperature, energy input, equal to the latent heat of fusion, is required to change the phase of the material, without increase in the melting material temperature. Hence, the phase change can be used to limit or delay the temperature rise in a system, much like evaporation. In the liquid or partially liquid regions, convective currents may develop, altering the temperature field, and in particular affecting the solidification process and the microstructures of the material (Faghri & Zhang, 2020).

Some studies that have utilized numerical modeling techniques to examine solidification and melting are listed here. Tabaara and Gu (Tabbara & Gu, 2012) have modeled the metallic droplet impingement, solidification and remelting, which can be applied to coating and other manufacturing/ surface treatment processes. El-Hadj et al. (El-Hadj, Zirari, & Bacha, 2010) have examined the deformation and solidification of copper droplets upon impact with a surface for thermal sprays. Stavroupoulos and Foteinopoulos (Stavropoulos & Foteinopoulos, 2018) have applied solidification/melting modeling to simulate the additive manufacturing processes. Florio (Florio, 2018) modeled the splat formation and solidification of copper drops impacting a surface. Selvnes et al. (Selvnes, Allouche, Sevault, & Hafner, 2018) have simulated the formation of ice and the melting of this ice associated with a refrigeration system for a food processing plant. These studies have shown the environmental, surface, any externally applied loads (mechanical/thermal) and stirring or agitation of the system can impact the phase change that develops.

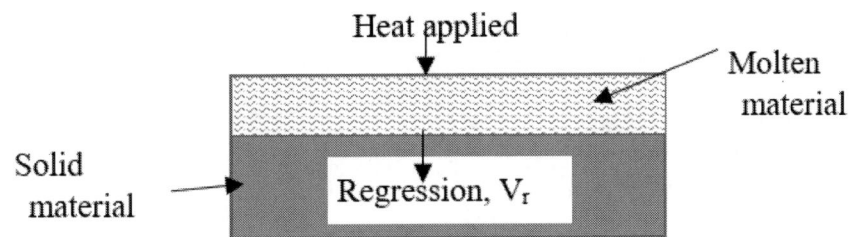

Figure 3. Melting/solidification phase change.

Two basic approaches are commonly used to model the solidification and melting processes. In both approaches, like with the evaporation and condensation, a VOF method is used to account for the presence of solid and liquid phases in the system. The first method is called an enthalpy-porosity method and was developed by Voller and Swaminathan (Voller & Swaminathan, 1987). This method does not explicitly trace the interface between the phases with time, but instead the method is used to indicate areas where the temperature falls between the liquidus temperature, T_L, and solidus temperature, T_S, of a material. Within this temperature range, the material is said to be undergoing a phase change with the liquid fraction, β. The region in which phase change is taking place is called the mushy zone.

$$\beta = 0; T < T_S$$
$$\beta = \frac{T-T_S}{T_L-T_S}; T_L \geq T \geq T_S$$
$$\beta = 1; T > T_L \qquad (26)$$

The latent heat content, ΔH^*, is used to account for the latent heat of fusion, so that for the material, the enthalpy, H, is given in Equation 27, where h is the sensible enthalpy and h_{if} is the latent heat of fusion for the substance.

$$H = h + \Delta H^*; \ \Delta H^* = \beta h_{if} \qquad (27)$$

The enthalpy/energy associated with the phase change is accounted for in this manner and the modified enthalpy formulation will essentially introduce an energy source term of $\dot{S}_e$ into the energy equation, since in addition to the sensible enthalpy the heat of fusion is added to the enthalpy value.

$$\dot{S}_e = -\frac{\partial}{\partial t}(\rho \Delta H^*) - \nabla \cdot (\rho v \Delta H^*) \qquad (28)$$

Volumetric sources are added to the momentum, $\dot{S}_{momentum}$, and turbulence governing equations to force the velocity of the solid material to zero. Therefore, with this method, any solid material will have a zero velocity, unless this source term is modified. The momentum source vector is given below where A_{mush} is called the mushy zone parameter and ε is assigned a small value to prevent division by zero and v is the velocity vector in the computational cell.

$$\dot{S}_{momentum} = -\frac{(1-\beta)^2}{\beta^3+\varepsilon} A_{mush} v \qquad (29)$$

A larger A_{mush} value therefore brings the solidifying fluid to rest more rapidly. For each of the turbulence parameters, if appropriate, a similar expression is used, but with the v vector replaced by the turbulence parameter scalar. Essentially, the porosity of the material decreases to zero as solidification occurs and increases to 1 as the melting process occurs.

With this enthalpy-porosity based method, the conservation of volume/mass may not be upheld, particularly as the actual volume changes with the phase change. Various means to account for potential net volume changes in the system studied due to the phase change have been devised. Dallaire and Gosselin (Dallaire & Gosselin, 2017) use the enthalpy porosity based method, but ensure global mass conservation by applying a moving wall to the container system studied for phase change. Again, however, since the enthalpy-porosity concept is used as the basis, the method does not explicitly track the interface between the phases. Since the interface is not explicitly defined, the resolution of the interface may be poor and so this may impact the simulation results.

A number of techniques based on moving interface tracking have also been developed as an alternative to the enthalpy-porosity based methods. Shyy (Shyy, 1994) developed a moving grid method with a change of coordinates set to conform to the interface shape. Other methods apply interface trackers, but are used with a fixed computational grid (Shyy, 1994). Ray et al. (Ray, Durst, & Kumar, 2007) present an algorithm where an Arbitrary Lagrangian-Eulerian type of model is developed with smoothing to reduce the skewness of the interface related cells. These methods require additional computational expense when compared to the simpler enthalpy porosity method, but may provide for more detailed results.

METHODS

With this foundational and background information for the common modeling methods, the development and application of modeling methods and techniques for phase change in particle/droplet flow conditions through the use of computational fluid dynamics is now described. All implementation of modeling methods and algorithms is done through the use of the commercial code FLUENT© with customization through user defined subroutines (UDFs). The methods developed and used for each of the phase change phenomena with particles/droplets are discussed below along with a definition of the problem used to establish the usefulness of the given method.

Ablative Related Phase Change

In this work, a method for modeling ablative related phase change from a particle surface is introduced. The method is based on the basic concepts of the general ablative modeling methods described earlier. When a particle is exposed to high temperature gas flow, the particle surface temperature increases. For certain materials or coatings on the particles, ablation phase change may occur. To capture the effects of the ablative process at the particle surface, the technique includes the following components:

1. The computational cells on the boundary of the particles discretized within the computational domain must be identified, both on the fluid side and the solid side of the particle.
2. The net mass flux of the carbon, $\dot{m}_c$, at the boundary of each computational cell on the fluid side of the particle-gas interface must be identified and the mass fluxes for the CO, $\dot{m}_{CO}$, O, $\dot{m}_O$, O_2, $\dot{m}_{O_2}$, and N, $\dot{m}_N$, gases from equations 4, 15, 16, 17, and 18 are calculated. The mass fractions of the gas in the computational cell bounding the wall, the local wall temperature, and local wall pressure are used in determining the parameters in the expressions including the mass fractions of the gas species. The density of the gas mixture in the bounding computational cell is used in these expressions as well as ρ_g, the density of the gas produced
3. With the rate of the carbon loss known from Equation 4, the rate of regression of the surface of the particle in the negative direction to the outward facing normal to the particle surface, V_r, can be determined from Equation 13. The regression rate at each fluid and solid computational cell face around the particle boundary is determined. At each node (on the fluid and solid side), an average

of the velocities from each of the connected boundary faces is computed. Then, over the computational time step, the displacement of the nodes in each coordinate direction is determined. See Figure 4. At Surface A and Surface B, the corresponding nodes are A_1 and A_2 for surface A, with the centroid at A_C, and B_1 and B_2 for surface B, with the centroid at B_C. The outward directed unit normals at the particle surface, directed from inside the particle to the gas outside, are given by $\hat{n}_A$ and $\hat{n}_B$. The magnitude of the regression velocities are V_{rA} and V_{rB}. Then, the velocity vectors for the regression along the two bounding surfaces are given in Equations 30 and 31.

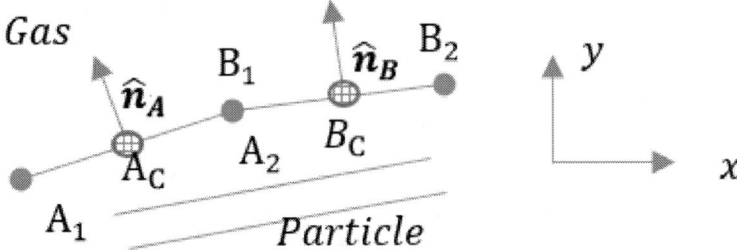

Figure 4. Geometry for specifying node motion.

$$V_{rA} = -V_{rA}\hat{n}_A \tag{30}$$

$$V_{rB} = -V_{rB}\hat{n}_B \tag{31}$$

The nodal displacement at the node A_2 or B_1 can then be written in the x direction and in the y direction. Each boundary node is assigned a displacement in this manner.

$$\Delta x_{A_2} = \Delta x_{B_1} = \frac{1}{2}\{(V_{rA} + V_{rB}) \cdot \hat{i}\} \tag{32}$$

$$\Delta y_{A_2} = \Delta y_{B_1} = \frac{1}{2}\{(V_{rA} + V_{rB}) \cdot \hat{j}\} \tag{33}$$

4. The node motion specified is then applied through a user subroutine so that the FLUENT© code can move the boundary nodes by the given amount and the geometry of the system can be updated. The frequency of this update can be modified to balance computations with accuracy, depending on the conditions in the system.
5. Consistent with the mass flux values calculated in step 2, the velocity of the net gas motion, V_g, into the gas computational cell next to the wall is tabulated using Equation 14. The density of the gas, ρ_g, used in this calculation was tested with both the density of the gas in the computational cell next to the wall and the density of the net incoming gas at the wall temperature and pressure. The second option requires the calculation of the mass fractions of each of the species in the mixture and so has some additional computational expense. As noted, the results reported use the density of the gas currently in the cell.
6. With this information, several volumetric sources can be specified to represent the effects of the net introduction of the gases from the ablation. The net volumetric mass generation, $\dot{m}_{Vc}$, and the net volumetric mass generation of species i, $\dot{m}_{Vi}$, flux based generation, is reformulated in the manner below where V_{cell} is the volume of the cell and A_{wall} is the surface area:

$$\dot{m}_{V_i} = \frac{\dot{m}_i A_{wall}}{V_{cell}} \tag{34}$$

A momentum source in each direction is added based on the normal direction to the wall surface, the velocity of the gas, and the mass generation rate as below.

$$\dot{S}_{V_M} = \dot{m}_{V_g} V_g \hat{n} \tag{35}$$

The volumetric energy source is equal to:

$$\dot{S}_{Q_M} = \dot{m}_{V_g} \sum_{i=0}^{Nspecies-1} Y_i \left(h_{ref, 298.15\,K} + h(T) - h_{ref, 298.15\,K} \right) \tag{36}$$

7. Once gases have been introduced into the main gas flow outside of the particles, chemical reactions are allowed to occur within the mixture of gas species. The chemical reaction set of Gupta et al. (Gupta, Yos, Thompson, & Lee, 1990)in Table 1 is used in this work.

To test the modeling implementation, a four, fixed particle system held within a channel is studied as shown in Figure 5. A large tank upstream of the channel initially contains high temperature and pressure gas that feeds the gas flow over the particles and initiates the ablative process. These four fixed 1mm radius particles are placed into the channel of a width of 5mm and a length of 82.5mm. The centers of the first column of particles is at 5mm from the entrance of the channel and centered at 1.25mm from the center of the channel. The gas initially in the tank is at a pressure of P_t and a

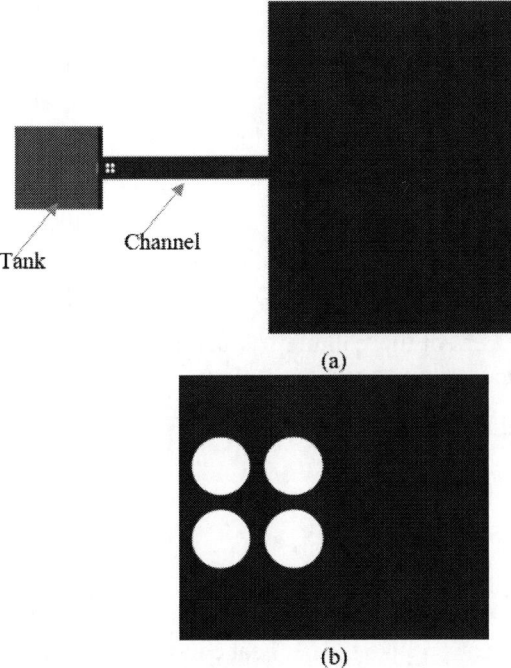

Figure 5. Ablation modeling system: (a) Entire system with tank, channel, and external (b) Four particles.

Temperature T_t. The simulation commences as a high speed flow develops and moves over the array of the particles. As the high temperature gas flows over the particles, the ablative pyrolysis gases are introduced into the gas zone just outside of the particle surfaces. The surface of the solid particles regresses by a consistent amount. Three initial tank pressures of 30MPa, 45 MPa, and 90 MPa all with an initial temperature of 2500K are assigned. The gas in the tank is composed of the following mass fractions: $0.78 N_2$ and $0.22 O_2$. In this way, any CO or CO_2 gases found within the system can be traced solely to the ablative process and the gas generation from the ablation can be clearly seen. Chemical reactions in the gas flow are free to take place as the gas species mix and interact.

Evaporation Phase Change

For the evaporative phase change modeling, a VOF based modeling method is used. No grid-fitted interface between the liquid and gas phases is implemented. Particularly with the evaporation/condensation of droplet, the large change in the droplet geometry due to the deformation associated with the high speed flow and due to the phase change process, produces conditions for which the mesh motion cannot feasibly capture the results due to the deformation. Instead, the interface between the phases is defined at the cells for which the volume fraction of the gas or liquid phases is not equal to one or the gradient of the volume fraction is non-zero. The unit normal of this interface points in the direction of the changes in the liquid volume fraction, α_l, for this case.

$$\hat{n} = \frac{\nabla \alpha_l}{|\nabla \alpha_l|}$$

The non-normalized gradient vector represents the interface area. While a number of publications use the net heat conduction in this normal direction as a measure of the heat input into a computational cell for phase change, in higher speed flow conditions, this may not represent the heat flow. For this work, the thermal storage or specific heat, c_p, multiplied by the desired temperature difference toward the phase change temperature is used in the formulation of the mass of liquid that evaporates, or conversely for the mass of vapor that condenses. The mass transfer or exchange between the phases is equal in magnitude, but opposite in sign. Here, the energy exchange is assumed to be only that required for the phase change or the rate of mass transfer multiplied by the heat of vaporization.

For the droplet evaporation/condensation conditions considered, the following components are included in the computational fluid dynamics modeling technique.

1. Two separate fluid "materials" are defined in the model; one for the liquid phase and one for the vapor phase. The appropriate reference enthalpies must be specified. The latent heat of vaporization, h_{fg}, must be included in the reference enthalpy of the vapor phase.
2. The surface tension effects are included in the model since the surface tension will impact the deformation of the droplets, the interface area, and therefore the flow and thermal conditions at the interfaces and throughout the system and the phase change that develops. The surface tension forces are calculated through the method devised by Brackbrill (Brackbrill, Kothe, & Zemach, 1992) mentioned earlier. Additionally, the surface tension at the walls of the channel are also included in the simulations with the input of a contact angle. The droplet shape at the wall/gas interface is influenced by the acting forces and is consistent with the applied contact angle.
3. At each of the computational cells, if a mixture of the gas and liquid is detected, indicating a phase interface, conditions are checked for potential phase change. If the saturation temperature at the partial pressure of the liquid has been reached, within a tolerance, then phase change is defined to be occurring. For liquid with temperatures at or above the saturation temperature plus a tolerance, evaporation is occurring and for a gas phase with temperatures at or below the saturation temperature minus a tolerance, condensation is

occurring. Figure 6 is a plot of the saturation temperature vs. saturation pressure for water.

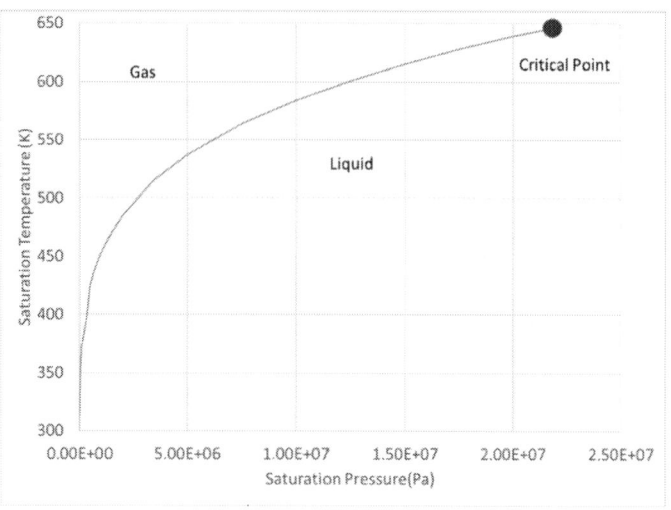

Figure 6. Saturation pressure and temperature for water from data in (Nellis & Klein, Thermodynamics, 2012).

4. If the phase change is taking place, the rate of mass consumption of one phase and the rate of mass generation for the other phase need to be determined. In Equation 37, the volumetric mass generation of the vapor phase $\dot{m}_{evap}$ is determined, but a similar approach can be used for the determination of the rate of condensation. The factor, f, is adjusted to maintain the temperature at the interface near the saturation temperature. A saturation temperature tolerance of 3K was used in this work with the intent on preventing quick changes between the phases for slight changes in the temperature. In this work a factor f of 100 was used. Vol is the computational cell volume, T is the cell centroid temperature, T_{SAT} is the saturation temperature, and Δt is the computational time step. For evaporation, the corresponding liquid mass generation rate is $\dot{m}_l$

$$\dot{m}_{evap} = f \frac{\left(\alpha_l \rho_l Vol \, c_{p,l}(T-T_{SAT})\right)}{\Delta t \, Vol \, h_{fg}} \tag{37}$$

$$\dot{m}_l = -\dot{m}_{evap}$$

5. Coupled to the calculation of the amount of gas evaporated, a set of mass, momentum, and energy sources can be specified to simulate the effects of the phase change process. The mass generation of the gas phase and liquid phase from Eq. 37 are applied to a given cell. If, for instance a phase, like a gas phase consists of multiple gas species, then not only does a net mass source of the phase have to be specified, but the same mass source is applied to the specific species of the gas phase as well. If the gas consists of N_2, O_2, and H_2O, then for water vapor evaporation, the mass generation of the vapor H_2O must be assigned. If the liquid phase for evaporation is moving, mass conservation at the interface can be approximated as:

$$\rho_g V_g = \rho_l V_l \tag{38}$$

A momentum source vector can be specified that uses the liquid momentum as a sink and the gas momentum as a source for evaporation:

$$S = \dot{m}_{evap}(V_g - V_l)\hat{n} \tag{39}$$

An energy source is also required. The volumetric energy source is equivalent to the latent energy as:

$$\dot{Q}_{evap} = h_{fg}\dot{m}_{evap} \tag{40}$$

6. Chemical reactions between the gas species are modeled.

In this work, a system consisting of two liquid drops initially within the channel in the tank/channel system used for the ablation modeling (without the four particles) (Figure 7) is used to implement and test the evaporation/condensation modeling methods. The channel dimensions are the same as those for the ablation model system in Figure 5. The high pressure and temperature gas in the tank drives the fluid flow over the two, initially circular droplets (Figure 7). The interaction of the droplets with the flow develops, leading to both the droplet deformation and motion and the evaporation from the droplet-gas interfaces. Two water droplets of 1mm radius are placed about 12.5 mm from the entrance of the channel, with a 2.0mm spacing between the centers of the initial drops. The phase contour plot shows the initial drop positioning. The high pressure and temperature tank gas is released as the simulations start with a 2500K initial temperature and a 90MPa and 110MPa initial tank pressure specified for two cases. The tank gas composition is a mass fraction of 0.78 for N_2 and 0.22 for O_2.

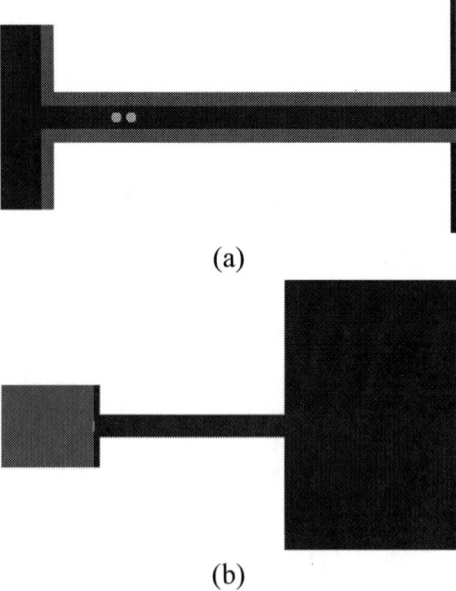

Figure 7. System considered (a) phase contours with liquid drop shown (b) overall system with high pressure/temperature tank on the left.

Water is used as the phase changing fluid and no water or hydrogen is initially in the gas phase in the system. The only reaction specified is the water vapor decomposition into hydrogen and oxygen, though additional reactions can be added as necessary. Hence, the only water vapor and hydrogen gas is due to the evaporation. The latent heat of vaporization of the water is set to 2.444e+06 J/kg. The appropriate reference enthalpies for the liquid and gas water materials are also assigned. The density of the liquid is set to a constant value of 998 kg/m^3 and the specific heat to a value of 4182.0J/kg-K. Ideal gas conditions are assigned to the gas mixture with the water vapor given a specific heat of 2140.0J/kg-K. The surface tension between the liquid water and the gas is assigned a value of 0.1N/m with a 90 degree contact angle. This system is used to demonstrate the evaporation/ condensation techniques developed and to highlight some of the important associated phenomena and information that can be obtained from the models.

Melting and Solidification Phase Change

The final phase change type for which the modeling approach developed needs to be explained is melting and solidification. A VOF based method combined with the solid particle boundaries is used to model the melting and solidification, building upon the evaporation/condensation modeling method described and the ablative modeling. The solidification/melting of interest for this work is that occurring near or at the surfaces of the particles. Hence, the addition or subtraction of solid material from the surfaces can be accounted for through the node motion on the surfaces of the particles, similar to the ablation modeling method. When the conditions are such that a phase change is occurring, the particle wall is moved so that when melting occurs, the melt is applied outside of the particle wall and the wall surface recedes. The liquid outside of the particle can move, deform, and interact with the flow, particles, or other objects in the system. If the solidification/ melting is occurring within the fluid, then the method follows that from the evaporation and condensation modeling method as defined earlier in this chapter. In order to model the melting/solidification phenomena, the following considerations are made for each of these two conditions – phase change along a surface and phase change within the fluid.

1. Two separate "fluids" are specified for the two phases of a given material 1- the solid and the liquid phases. The viscosity of the solid is set to a large value to mimic the rigid motion the solids should have.
2. As with the evaporation/condensation, the amount of mass undergoing a phase change has to be calculated. The determination of the mass undergoing a phase change closely follows the evaporation/condensation method from the previous section, but with the T_{SAT} now set equal to the melt temperature and the h_{fg} set to h_{if}, the latent heat of fusion instead of the latent heat of vaporization in Equation 37. For melting, $\dot{m}_{melt}$, is the volumetric rate of melt or liquid phase generation and the corresponding change in the solid mass, $\dot{m}_s$, can be found. Adjustments to this method can be made for the solidification process.

$$\dot{m}_{melt} = f \frac{\left(\alpha_s \rho_s Vol\, c_{p,s}(T-T_{SAT})\right)}{\Delta t\, Vol\, h_{if}} \tag{41}$$

$$\dot{m}_s = -\dot{m}_{melt}$$

3. Based on the mass melted, if the melting is occurring near the surface of a particle/wall, the surface regression velocity, V_r, at the particle surface can be determined, following the method used in the ablation model, where in Equation 42 s is for solid and Vol is the computational cell volume.

$$V_r = \frac{\dot{m}_{melt} Vol}{\rho_s} \tag{42}$$

Then, the liquid velocity at the interface, V_l, can be calculated as:

$$V_l = \frac{\rho_s V_r}{\rho_l} \tag{43}$$

If the melting and solidification is occurring in the main flow, not along a surface, the same calculations can be made to obtain the phase velocities.

4. As with the evaporation/condensation, a set of sources needs to be specified to model the phase change. When phase change is occurring at a solid/particle surface, the mesh motion/surface regression accounts for the mass change and no mass sources are applied to the solid and only the liquid source in Equation 41 is applied. If the phase change is occurring within the general open fluid, both solid and liquid mass sources are applied in accordance with the rates in Equation 41. Unlike the enthalpy-porosity model described in the background where momentum sources are applied specifically to force the absolute velocity of any solid components to zero, in this modeling method, only the momentum due to the phase change may be considered. If a droplet in a high speed gas flow "solidifies," that material must be allowed to continue to move within the gas based on the fluid induced forces. So, for the conditions where the melting is occurring at the wall of an object/particle, the momentum source on the fluid side of the fluid-solid particle wall interface is:

$$\dot{S}_{V_M} = \dot{m}_{melt}(V_l)\hat{n} \qquad (44)$$

where $\hat{n}$ is the outward directed normal from the particle surface. The volumetric energy source associated with the melt introduction is then:

$$\dot{Q}_{melt} = h_{total,melt}\dot{m}_{melt} \qquad (45)$$

$h_{total,melt}$ includes the formation and latent heat enthalpies. If the melting is occurring in the flow alone, then:

$$\dot{S}_{V_M} = \dot{m}_{melt}(V_l - V_s)\hat{n} \qquad (46)$$

where $\hat{n}$ is the interface normal, just like in the evaporation/condensation models. The energy source associated with the melting is then:

$$\dot{Q}_{melt} = h_{if} \dot{m}_{melt} \tag{47}$$

5. Modifications can also be made to allow for a melt temperature and related fluid properties that vary with the local pressure or solid properties that vary with temperature. In the current implementation, the melt temperature and the property values are held constant.

The melting/solidification modeling method is implemented and tested using the geometry of the ablation test case in Figure 4. However, for this work, instead of gas, the region outside of the particles is filled with a liquid material 3K above the saturation or melt temperature, just outside of the tolerance. Four particles are held in the channel and the particles remain fixed, but for any surface regression. The material involved is copper with a melt point at 1356K. The solid particles are set to temperatures of 100K, 200K, and 300K below the melting point. Solidification develops at the outer surface of the particle and spreads through portions of the liquid. The different levels of solidification with the various initial particle temperatures can be examined. Also of note, particularly with the melting and solidification, with the relatively large densities and different densities between the phases, the net volume/mass is not strictly conserved. With the small amount of material that changes phases for the cases considered, any issues with the conservation of volume for liquids and solids with different densities should be relatively small. Modifications to shift or alter cell volumes can be made in subsequent revisions of the modeling method.

Other Modeling Considerations

In all of the simulations for the implementation of the three major modeling methods, the commercial code FLUENT© is used. The coupled pressure based solver was used with second order in time discretization, bounded central differencing for the momentum equation, second order upwinding for the energy equation, species equation, and density equation discretization. Second order discretization is used for the pressure as well. Least squares cell based gradient discretization is implemented. For the gases, kinetic theory is implemented for the calculation of the thermal conductivity and the viscosity of the components of the fluid and a piecewise polynomial for the specific heats. Ideal gas mixing is used for the formulation of the properties and for the equation of state of the mixture of the gases where a gas constant based on the mass fraction of the mixture is used. Kinetic theory is applied for the mass diffusivity in the mixture. For the melting and evaporation modeling, the VOF modeling method is selected. With the VOF modeling method, the continuum surface force method is implemented with a surface tension coefficient of 0.1 N/m between the liquid and gas phases in the models. An explicit VOF formulation is selected, as this proved the most stable. A sharp interface modeling method is selected with a velocity base sub-time step calculation method and the option to solve the VOF every iteration.

A fixed time step of approximately 5.0e-09s is implemented, with adjustments required for convergence for the various problems and cases considered. The simulations were all run on 72 nodes on a Linux HPC with approximate run time of 24 hours, with slightly longer times required for the VOF models. FLUENT 2021 R1 was utilized for these simulations.

A mesh sensitivity study was performed for the original ablation model configuration. Comparisons of the net amount of CO released from the particle surfaces as well as the net force acting on one of the upstream particles were used to measure when sufficient mesh resolution had been reached since these parameters would capture the effects of the ablation and the flow field. A quadrilateral mesh, three elements deep is applied to both the inner and outer surfaces of the particles where the resolution of the flow, temperature and pressure fields, the shear stresses, heat flux, and species flux are important. A triangular mesh is applied in the remainder of the system. The triangular mesh is required for the general motion of the particles through the flow and thus for the general regression of the surfaces of the particles. The time period covered in these simulations was not sufficient to cause a significant decrease in the radius of the particles. The current modeling method allows the particle size to shrink drastically, but some finite dimension must be maintained. For the final mesh selected, 56 cells are located around the circumference of the circular particles with the nearest normal node to the wall at 1.00×10^{-5}m. The mesh near the particle walls is shown in Figure 8. The y+ values over the transient flow conditions studied were less than 10. Further refinement in the mesh should be examined. However, for the purposes of exploring the modeling methods, in the interest of computational resources, the mesh as stated was used. The complete system has a mesh size of about 600,000 cells.

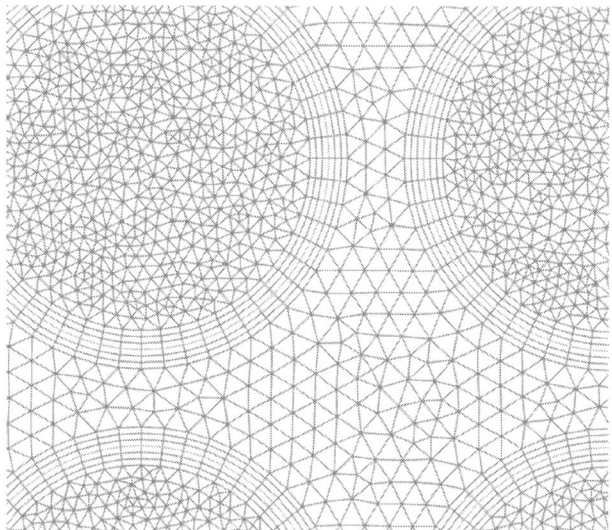

Figure 8. Mesh in the vicinity of the particles.

RESULTS

The simulation techniques and methods described are implemented for the specific systems and parameter cases defined. The aim of these limited studies is to test the basic functioning of the models, gain insight into the three phase change phenomena, and to establish the potential usefulness of the modeling methods. The results for the ablative phase change, evaporative phase change, and solidification phase change are presented and discussed in the following sections.

Ablative Related Phase Change Results

For the ablation modeling implementation, the velocity, temperature, pressure, and CO_2 gas mass fraction are compared as the flow develops with time amongst the different initial pressure cases. First, however, the effect of the turbulence model on the simulation results is briefly discussed. The κ–ε realizable turbulence model was initially selected for these simulations. A DES model with a κ–ω RANS model was also tested for the tank case of an initial 90MPa pressure level. The pressure field and the CO mass fraction at the same time as well as a velocity vector plot are shown in Figure 9. Clearly the results show differences in the flow patterns and CO distributions. The results with the DES model appear to better match the expected diamond shock structure that should develop in the channel, the CO distribution with the gas patterns following the eddies or circulation regions in the flow, rather than being uniformly dispersed through the channel, and the more robust circulations upstream, between, and forming downstream of the particles. These results are also consistent with the expectations for the types of flow patterns between the particles as per the flow over cylinders reported in Lam et al. (Lam, Gong, & So, 2008). The trends in the flow patterns follow the attributes of the results of simulations using a RANS model as opposed to those using an LES model (Jiang & Lai, 2009). Therefore, for the remainder of this work, the DES turbulence model is utilized to ensure consistency across all of the simulations conducted in this chapter.

Now, the conditions that develop as the gases move over the particles can be described and the effect of the tank pressure on the system conditions can be examined. As the hot gases from the tank impinge upon the fixed particles surfaces, the gases are slowed and the gas temperatures increase. The ablative gas generation commences and changes based on the local near-wall gas conditions develop, including the gas temperatures, velocities, pressures, and composition. The general velocity field conditions show a diamond-like structure forming in the channel (Figures 10, 11). This structure is due to shock/pressure wave formation as the gases move from the tank into the channel and then the interactions of these shocks with the channel flow path surfaces. The strength of the diamond patterns appears to diminish with time/distance from the flow channel entrance. The vector plots also show the circulation regions that develop near the particle surfaces. The circulations are more vigorous for the higher tank pressure cases, in part because the fluid has been moving over the particle surfaces for a longer period of time due to the faster fluid flow.

The patterns in the temperature field follow those in the velocity field with the higher temperature regions occurring near the particle surfaces and the effects of the flow modulation in the velocity field reflected in the diamond patterns in the temperature field (Figure 12). A thinner region of high temperatures develops near the walls of the channel where the flow velocities are naturally lower due to the presence of the fixed wall and where viscous effects are more prominent as the flow velocities reduce to zero. Overall, the peak velocities and temperatures are higher with the higher initial tank pressure due to the greater energy held by this gas feeding the operation of the system.

Because the higher pressure cases result in faster moving flows, the locations of the "front" of the tank gas flow moving along the channel differ with the tank gas pressure as do the velocity, temperature, and pressure magnitudes. At the time in these plots, the flow has only been recently moving over the particles for the 30MPa case, but has been moving over the particles and is reaching the end of the flow channel for the 60MPa case. The visualization of the local conditions helps to explain trends and patterns and can be coupled to the ablative gas conditions that develop.

The CO distribution can be used to indicate the proper operation of the ablative model. The CO is the only carbon based gas species generated by the model of the ablation at the particle surfaces and in this simulation no other carbon based gas species is initially included in the model. Contours of the CO distribution at the same time for the three different pressure distributions are provided in Figure 13. The local distribution of the CO varies along the circumference of the particles and the characteristics of this distribution are different for the leading two particles compared with the trailing two particles. The flow conditions and temperature conditions, observable from the previous vector and contour plots, indicate the underlying reason for the distribution. The upstream particles tend to have the higher mass fraction of CO through about forty five and sixty degrees counter-clockwise from the downstream channel direction and symmetrically through about forty five and sixty degrees clockwise. For the second column of particles further downstream, the highest CO mass fraction occurs just near the downstream zero degree angle with the channel axis. The circulation zones, particularly those behind or downstream of the particles can influence the local CO distribution with the circulations moving the CO away from the surface, or trapping specific gas species near the surface. High temperatures are also maintained in these circulation regions, contributing to the higher CO concentration. The local circulation patterns, thus lead to the local CO patterns near the walls of the particles. (Figures 19-12) The CO carried downstream follows the eddies and circulations of the flow along the channel, leading to the CO mass fraction patterns in Figure 13.

Specific data can be retrieved from the models. Figure 14 shows the volumetric generation of CO at 1e-05s and 8e-05s for the 60MPa case, indicating the generation distribution becomes more uniform with time as the temperatures around the particle become more uniform and the areas with the higher temperature tend to have a higher rate of production. The composition of the gas near the wall also impacts this generation rate and so as the temperature/property dependent reactions occur in the nearby gas species, the CO generation rates are altered. The CO mass fractions along the upper left particle circumference for the three pressures at the same time at 1.0e-05s and 8.0e-05s are shown in Figure 15. The areas with higher temperature tend to have the higher rates of generation and the generation tends to become more uniform over time as the heating of the particle becomes more uniform. Note in these plots, the curve length begins at a negative 30 degree angle from the downstream side of the upper left particle and then moves counter-clockwise around the particle circumference.

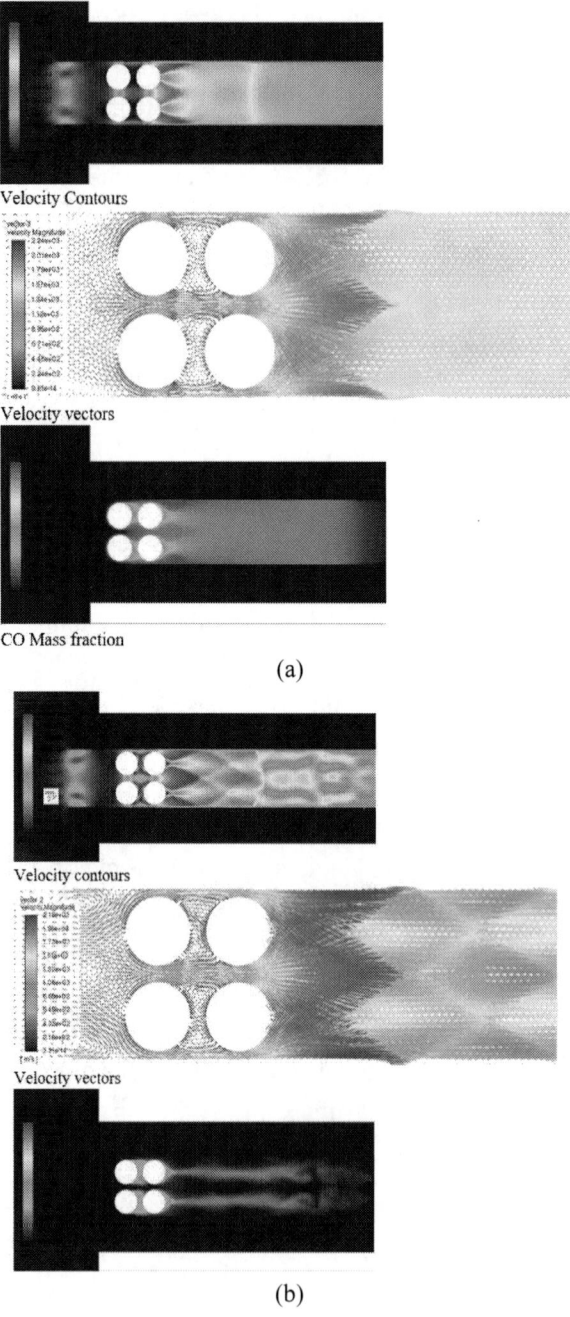

Figure 9. Condition comparison turbulence model variation for the ablation model (a) $\kappa-\varepsilon$ model (b) DES model.

Phase Change Simulations in Particle Flow ... 37

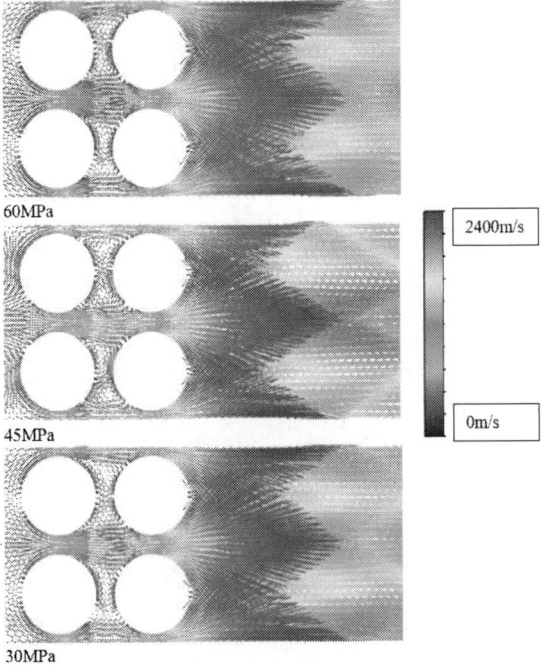

Figure 10. Velocity contours at tank pressures indicated at 8.0e-05s.

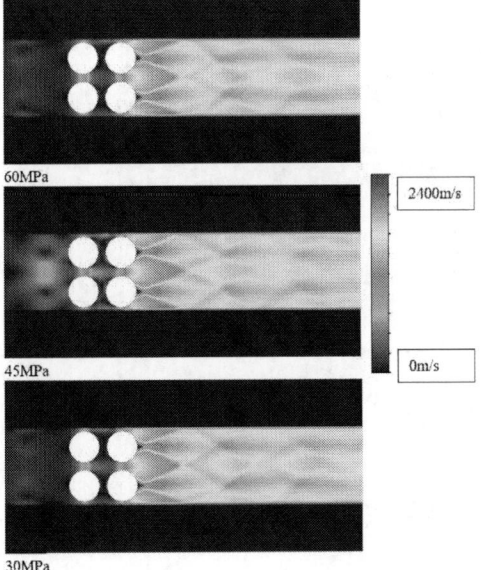

Figure 11. Velocity contours at tank pressures indicated at 8.0e-05s.

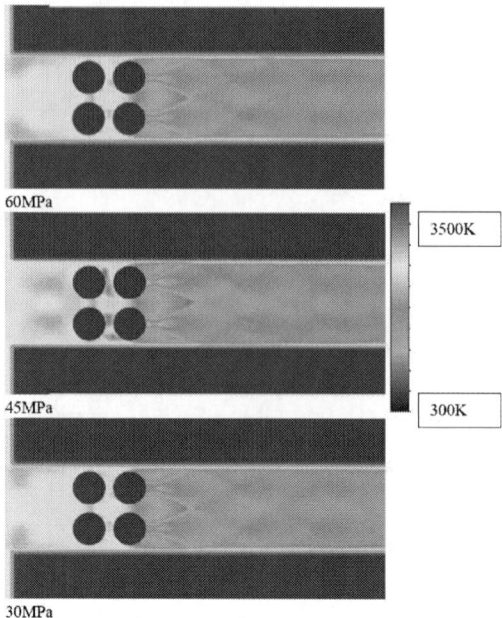

Figure 12. Temperature contours at tank pressures indicated at 8.0e-05s.

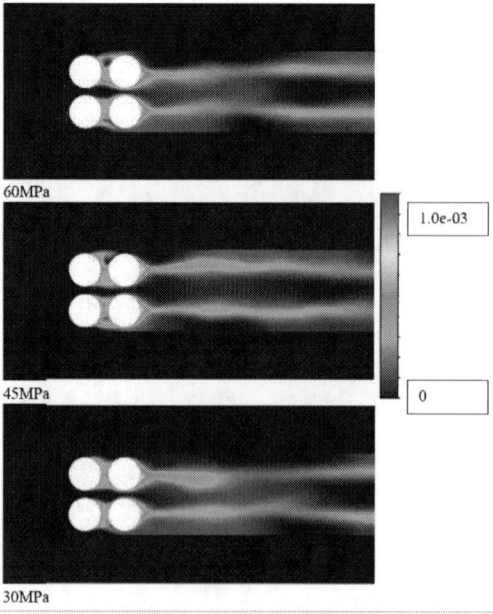

Figure 13. CO Mass Fraction contours at tank pressures indicated at 8.0e-05s.

Phase Change Simulations in Particle Flow ... 39

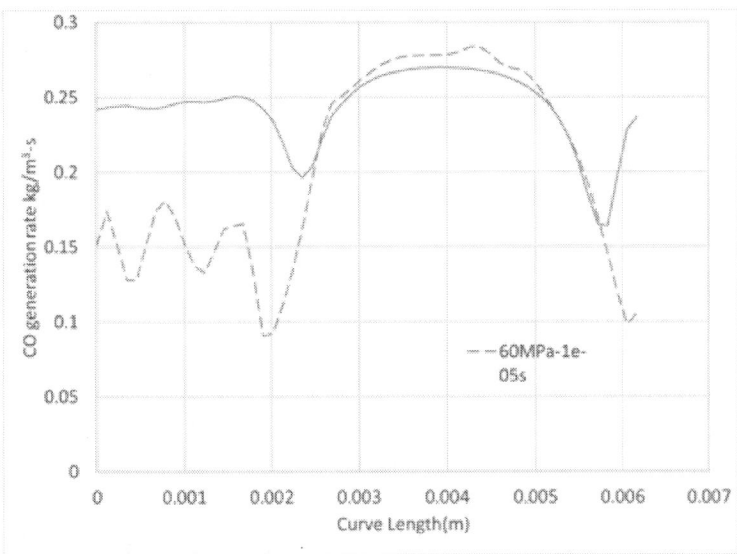

Figure 14. CO volumetric generation rate around the circumference of the upper left particle.

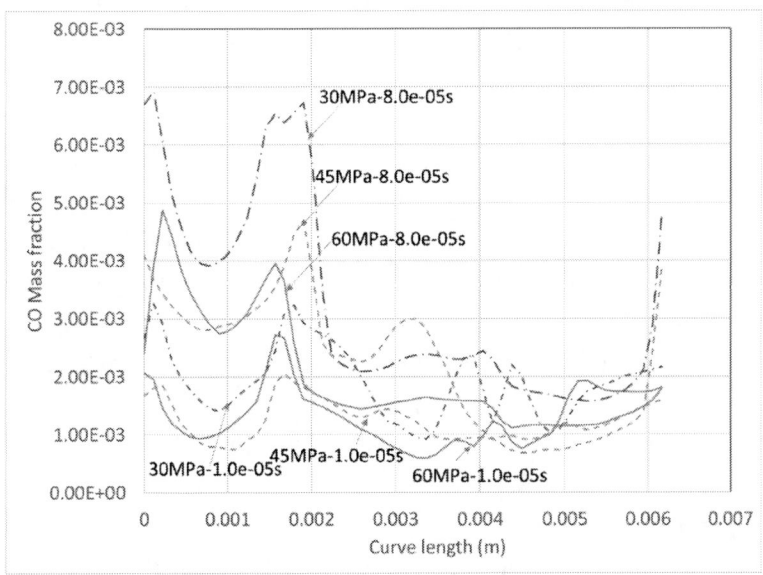

Figure 15. CO mass fraction around the circumference of the upper left particle for conditions indicated.

The results of these studies have clearly shown the utility of the modeling method. The local conditions can be used to determine the local rates of ablative gas generation. The gas flow and subsequent reactions in the gas can be included in the models. In addition to specific data, visualization of the transient flow features, temperature contours, and gas species distributions provide insight that is difficult to obtain through solely physical testing. Hence, this modeling method allows for the estimation of detailed conditions within and around the particles and in the resulting gas flow moving along the remainder of the channel.

Evaporation Phase Change Results

The next set of studies conducted involves the evaporation phase change modeling method. The simulation involves deformation, motion, and breakup of two water droplets (Figure 7), the production and transport of the water vapor that results from the evaporation from the droplet surfaces, and the changes in the typical flow and temperature fields that result from the effects of the droplets. No water vapor or hydrogen gas is introduced into the system initially so the water vapor present in the system is due to the result of the vaporization of the original water droplets. The results can be examined to provide a better understanding of the phenomena that develop during droplet evaporation in high speed compressible flow conditions.

As the high speed, high pressure, and high temperature gas begins to interact with the dense liquid droplets, the gases are slowed, much like the gas-particle interactions in the ablative modeling. However, unlike the particles, the droplets deform as a result of the fluid forces acting upon the droplets. A series of contours in Figure 16 show the severe droplet deformation for the 90MPa tank pressure case. While the droplet is deforming, the gases are also working to heat the liquid at the gas-droplet interface. When a sufficient temperature is reached for the local pressure conditions, phase change from the liquid to the vapor at the interface region develops. In tandem with the deformation and evaporation, the droplets are moving downstream under the effects of the net fluid induced forces acting on the droplet. Depending on the local shear, pressure, and surface tension effects, portions of the droplet may become thin and break away from the original drop, forming smaller droplets. When some of the liquid material contacts the surface of the channel, near a region of slower flow, the liquid material may "stick" to the channel surfaces for some time depending on the local forces and the surface tension and contact effects at the fluid-channel wall interface. These conditions can be clearly seen in the phase contour plots and as smaller and more numerous droplets progress further downstream and an agglomerate of the liquid material remains in the low velocity zone near the channel wall.

The density, viscosity, thermal conductivity and other properties of the droplet are significantly different from the properties of the gas mixture in the channel. Hence, the local distribution of the droplets through the system, in turn impacts the flow, pressure, and temperature field that develops. Flow may be diverted around the heavier, slower moving liquid material. The low velocity, high temperature, and high pressure zones that form just as the flow begins to impact the droplets can be seen in Figures 17-20. Pressure waves/reflections develop within the flow channel. These can clearly be seen in the zones marked with an arrow in Figures 17 (the velocity) and 18 (the temperature) noted by the backward moving lower velocity zone. The effects of the flow blockage due to the initial presence of the drops remain long after the droplet has changed shape and the majority of the droplet has evaporated.

Initially the upstream side of the upstream droplet is "flattened" by the gas flow, shielding the downstream droplet for some time.(Figure 16) Though due to the nature of the flow, the deformation is more intense, the general patterns of the droplet shape follow those reviewed in Michaelides (Michaelides, 2006). The highest rates of vapor production initially come from this upstream side where the gas/interface temperatures are highest. The initial vapor plume sweeping around the first droplet and over the second can be seen in Figure 20. The significant pressure rise in the channel due to the obstruction caused by the droplets in their initial configuration can also be observed as in Figure 19, specifically at 4.0e-05s indicated by the arrow. Companion to the high pressures are the high temperatures and low velocities (Figure 17 and 18). The high pressures tend to lead to significant deformation of the liquid material. The two drops eventually coalesce under these high forces and are thinned into a long extended shape, a shape that reduces the net force acting on the liquid material (Figure 16). As the droplet thins, portions of the material break away, and the high pressure region behind the droplet subsides. The droplet continues to break apart, with portions interacting with the gas streams and portions interacting with the walls. Even with the smaller droplets and droplets near the wall, the impact of the liquid material on the velocity field can be seen in Figure 17. The correlation between the positioning of the liquid and the pressure, temperature, and velocity fields is apparent in the contour plots and is an important consequence of having the liquid drops in the flow stream (Figure 16-19). The pressure waves/reflections introduced from the initial flow obstruction are still affecting the flow as the gas and droplets are exiting into the surroundings.

The progression in the droplet contours exemplifies the deformation and break-up of the droplets, but also shows the decreasing net volume of the liquid material, indicating a transition from the liquid to the vapor through the evaporation process. The more distributed nature of the droplets and the more motion and mixing of the gas flow lead to greater mixing of the vapor in the flow channel as can be seen as time progresses. The development of a large region of H_2O vapor in the channel is the result of the evaporation processes (Figure 20). The introduction of the vapor into the gas stream occurs at the liquid-gas interfaces, but the gas flow dictates how the vapor moves along the channel. Figure 21 provides the H_2O mass fraction along the channel centerline. The initial increase in the H_2O levels and the motion of the H_2O downstream can be seen, with the mass fraction remaining steady or decreasing after 8.0e-05s. Chemical reactions between this vapor and other gas species is considered, with a limited reaction in the current model. As the gas accelerates upon reaching the outlet of the channel, the presence of the liquid material in the flow stream is also apparent. The expansion process upon exit from the channel accelerates the gas and reduces the gas temperature, potentially leading to phase change from a vapor to a liquid again. This condensation was not considered in the current work but can be incorporated into the modeling method.

The results of the evaporative type models have shown that the simulation is capable of incorporating the droplet deformation and break-up, the liquid evaporation, the water vapor production and potential reactions with the gases and the interactions between the gases, the droplets, and the channel walls and other geometry. Hence, the modeling techniques can provide information about the interdependencies between the flow conditions, the droplet shapes, the rates of evaporation, and the distribution of the gas species in the system. Changes in the system operating conditions, geometry, drop size or position, or other aspects can be virtually tested as a means to tailor the multiphase flow for a specific purpose.

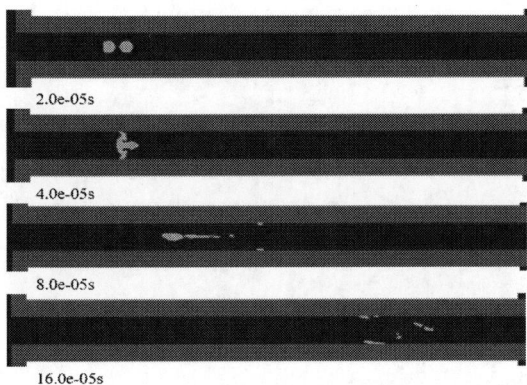

Figure 16. Evaporation model, fluid phase contours for 90MPa tank pressure: Red=channel wall, blue=gas, green=liquid.

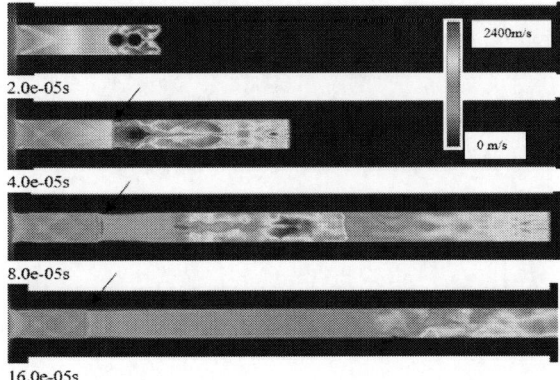

Figure 17. Evaporation model, velocity contours for 90MPa tank pressure.

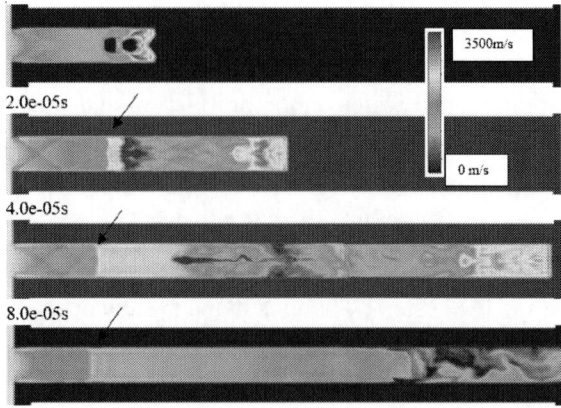

Figure 18. Evaporation model, temperature contours for 90MPa tank pressure.

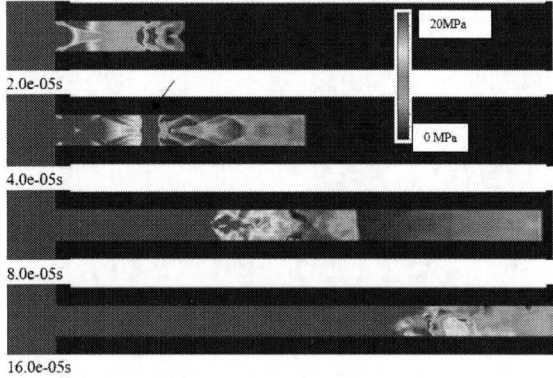

Figure 19. Evaporation model, pressure contours for 90MPa tank pressure.

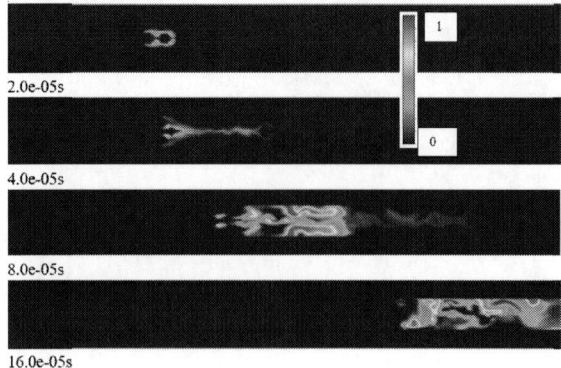

Figure 20. Evaporation model, H_2O mass fraction contours for 90MPa tank pressure.

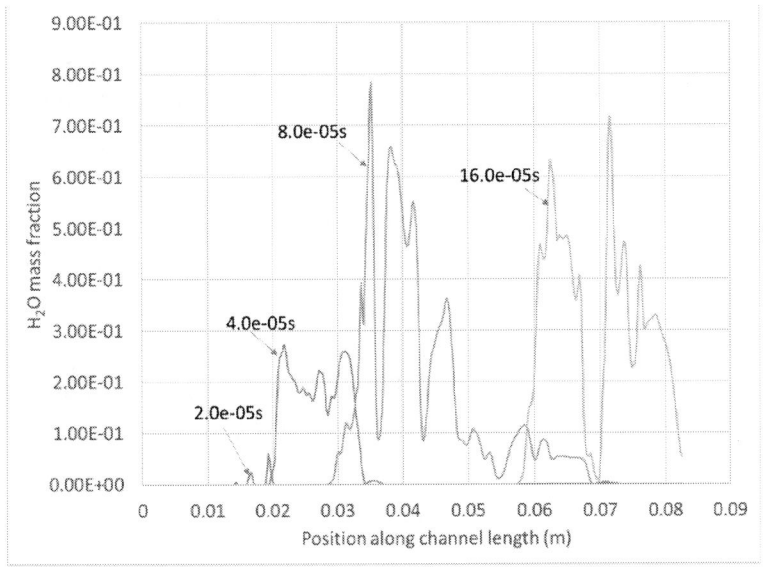

Figure 21. Evaporation model, H_2O mass fraction along channel centerline for 90MPa tank pressure.

Solidification Phase Change Results

The melting and solidification modeling method was implemented with the four particle system in Figure 4. The solid particles are initially set to a uniform temperature below the melt temperature to promote swift solidification in the surrounding initially liquid material. A range of initial temperatures is specified. No externally driven flow of the liquid over the particles is considered in this work. In addition to demonstrating the functioning of the modeling method, the basic trends in the development of solid material around the particles can be seen in the results of the simulations as well as the impact of the driving temperature difference between the liquid and the solid phases of the material.

The progression of the development of the solid material near the surface of the particles is captured in the modeling method. For all of the initial temperature cases, the solid material begins to build around the circumference of the particles in a uniform manner. The contour plots in

Figure 22 show the material that has changed phase in red for the case of the 300K temperature difference while Figure 23 shows the corresponding temperature field. The conditions for the lower temperature difference of 100K are provided in Figures 24 and 25 for the solidified material and for the temperature respectively. Over time, some variation in the distribution of the solid material build-up around the particles is noted. These variations are due in part to the local velocities that develop at the interface, though the velocities are under 0.01m/s. The circulating flow zones can be seen in Figure 26 and work to mix/alter the liquid streams near the wall, dragging in more of the higher temperature liquid or pulling in more of the cooler fluid near the wall. This circumferential variation becomes more pronounced over time and when the temperature difference driver is larger. Strut like features begin to build between the particles and the particles and the wall where the velocities are lowest and heat conduction is highest. These struts are mainly due to the heat conduction through the liquid and the lower liquid temperatures that then develop, leading to the solidification. The struts are generally found along paths with a "minimal" distance to the surface of the neighboring particle. Also of note is the significantly different extent of the strut development with variation in the initial temperatures in the liquid. Solid features extending between the particles only develop for the highest temperature difference case where the energy exchange is highest and solidification rates are therefore larger.

The changes in the local temperature distribution within the particles during the process can also be observed in the simulations. As the solidification occurs, the temperature around the circumference of the particles increases as a result of the heat transfer between the hotter liquid and the colder particle surface (Figures 23, 25). This energy is used to heat the particles, but as the temperature lowers in the liquid as a result, solidification of the liquid around the surface of the particles develops. The rate of the heating of the particles is greater for the higher temperature difference cases since a larger driver for the heating is present. The heating is relatively uniform around the circumference of each particle. For the higher heating rate/higher temperature difference cases, the thermal conditions are likely to become more asymmetric with time. The lower temperature regions that develop in local zones are related to the low velocity circulation zones around the circumference of the particles as well as the greater rates of heat conduction in the lower velocity zones with shorter distances between the low temperature particle surfaces.

The effect of the initial temperature difference or the heat flow potential/driver on the solidification can clearly be seen when comparing the results. The higher rate of heat transfer develops around the particles when the initial particle temperature is lower. The temperature of the liquid drops at a faster rate, leading to a greater rate and extent of solidification as seen comparing Figures 22 and 24. The lower temperatures that develop within the liquid near the particle surface gradually reach the phase change temperature and so the material begins to solidify in these zones. For the higher heating rate cases, the areas leading between the shortest distance to the nearest particle or the areas where the local velocity flow promotes lower temperatures reach the phase change temperature more quickly and over a larger area. The liquid begins to solidify at radii further from the walls of the particle at certain circumferential locations, particularly in the zones connecting particles where the nearby liquid is in closer proximity to other the cooler particles. Strut like patterns connecting the particles and connecting the particles to the walls develop (See Figure 22 and 24). Some small regions to the upstream and downstream side also show solidification away from the surface of the particle, a result of the lower velocity circulation zones transporting the cooler fluid further from particle surfaces. Such details can be obtained from the results of the simulations.

The melting/solidification modeling method provides for information on the thermal conditions and phase conditions that would be difficult to obtain solely through testing. The growth of the solidified regions with time, the development of small circulation zones near the surfaces of the particles, and the changes in the temperature field with time can be observed in the model results. The reason for the growth of the solid material between the particles can be gleaned from the local temperature distributions and the flow conditions. Therefore, the simulation technique provides benefits to understanding the solidification/melting phase change and improvements can be made to better model the events and to capture and to incorporate additional effects as needed.

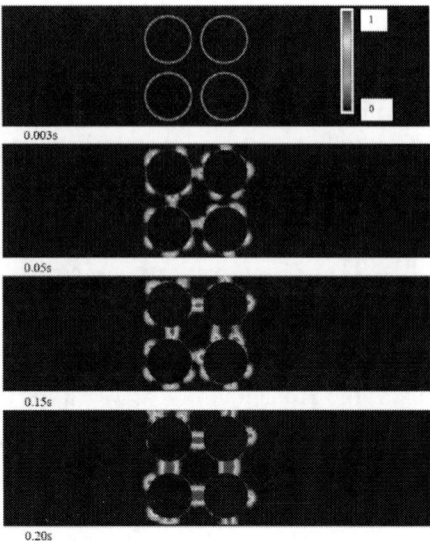

Figure 22. Solidification model, Phase changed material contours for 300K temperature difference.

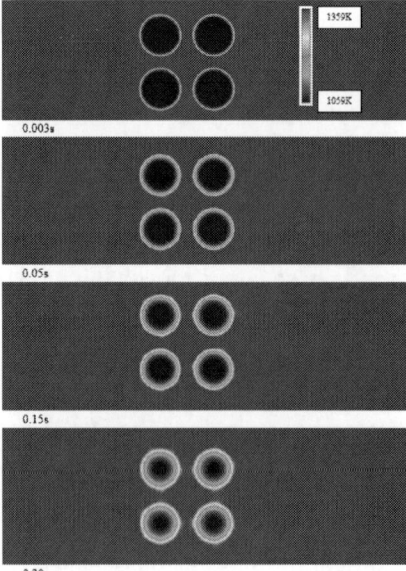

Figure 23. Solidification model, Temperature for 300K temperature difference.

Phase Change Simulations in Particle Flow ... 49

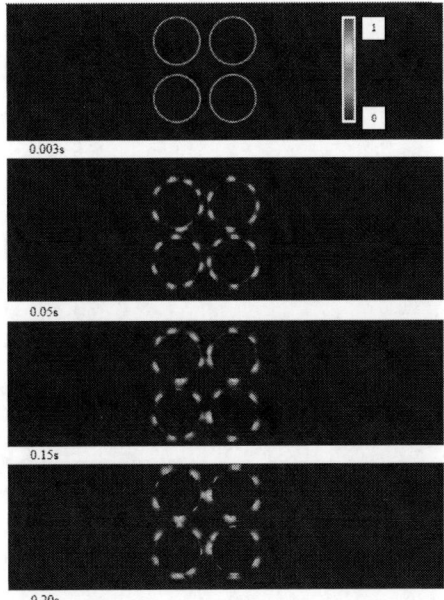

Figure 24. Solidification model, Phase changed material contours for 100K temperature difference.

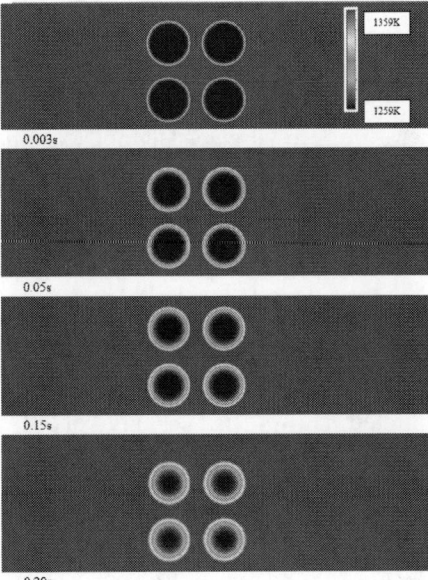

Figure 25. Solidification model, temperature material contours for 100K temperature difference.

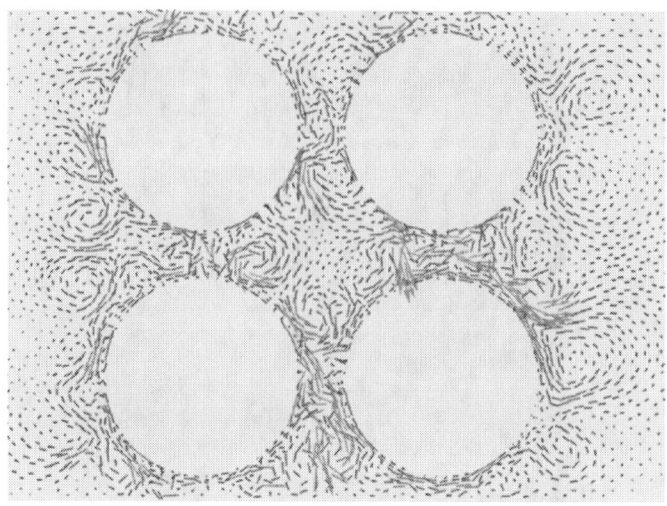

Figure 26. Solidification model, velocity vectors, 0.1s 300K temperature difference.

CONCLUSION

The work presented in this chapter describes a set of modeling methods that can be used to study phase change in systems, particularly fluid flow systems with particles and droplets. The computational techniques allow for the local flow, temperature, pressure, phase, and gas species conditions to be determined and used in finding forces, chemical reactions, or phase change. The effects of the particle and droplet positioning on the local flow, thermal, and species concentration conditions can readily be observed and are utilized in simulating the physical phenomena. These capabilities provide a means to obtain a deeper understanding of the phase change events that are occurring.

The study conducted shows that a number of new features and capabilities can be included in the simulations. First, for the ablation modeling, the motion of the particles can be incorporated into the simulations as the main intent of the work is to apply the methods to particle laden high speed compressible flow. Modifications to the regression model might need to be made to ensure conformity with the movement of the particles and with the energy balance at the fluid-gas interfaces. The ablation related gas release could be extended to different materials. The effects of the introduction of molten material due to the ablation could also be investigated.

For the evaporation modeling, a number of areas can be explored. Simultaneous evaporation and condensation occurring in the same system can be introduced so that while one region of the system may contain an evaporation event, another region in the system might be, independently, undergoing a condensation process.

Care should be taken not to allow for a quick switch between evaporation and condensation in a single phase changing computational cell, with the temperature tolerance a potential solution. A study of the effects of different means of calculating the mass involved in the phase change can also be pursued. The use of mesh motion can also be investigated as a means to ensure conservation of mass/volume.

Evaporation/condensation occurring on the surfaces of moving particles could also be studied under conditions where thermal and mechanical interaction develops between the particles and between the particles and the flow path.

For the solidification and melting, capability development can also be pursued. Similar to the evaporation/condensation, the simultaneous application of phase changes for both solidification and melting in the same system can be investigated. Modifications to the mesh motion for phase change at a wall or particle surface or in the general flow could be investigated to test for mass/volume conservation. The motion of the particles through the system, including mechanical and thermal interactions could be incorporated.

Overall, the methods devised for phase change reported here are customizable and can be updated/modified as new capabilities are required and as new phenomena need to be studied. The method offers the ability examine the conditions required for phase change and the conditions that develop during and after phase change with a high level of detail and under a controlled environment. The methods work to advance the state of the art in computational fluid dynamics simulations of phase change in particle flows.

REFERENCES

Alanyalioglu, Cetin Ozan. 2019. "*Numerical simulation of charring ablation coupled with computational fluid dynamics.*" Masters Thesis; Middle East Technical University.

Brackbrill, J. U., D. B. Kothe, and C. A. Zemach. 1992. "Continuum method for modeling surface tension." *Journal of Computational Physics*, 335-354.

Chen, Y. K., and F. S. Milos. 2004. "Finite rate ablation boundary conditions for a carbon-phenolic heat shield." *37th AIAA Thermophysics Conference*. AIAA. AIAA Paper 2004-2270.

Dallaire, J., and L. Gosselin. 2017. "Numerical modeling of solid-liquid phase change in a closed 2D cavity with density change, elastic wall, and natural convection." *International Journal of Heat and Mass Transfer* 903-914.

El-Hadj, A. A., M. Zirari, and N. Bacha. 2010. "Numerical analysis of the effects of the gas temperature splat formation during thermal spray process." *Applied Surface Science* 257: 1643-1648.

Faghri, Amir, and Yuwen Zhang. 2020. *Fundamentals of multiphase heat transfer and flow.* Switzerland: Springer Nature.

Florio, L. A. 2018. "Simulation of motion, deformation, break-up and deposition of copper droplet transported in internal compressible flow including phase change effects." *International Journal of Heat and Mass Transfer* 127: 658-676.

Gupta, R. N., J. M. Yos, R. A. Thompson, and K. P. Lee. 1990. A review of reaction rates and thermodynamic and transport properties for an 11-species air model for chemical and thermal non-equilibrium calculations to 30000K. *NASA Reference Publication 1232*.

Jiang, Xi, and Choi-Hong Lai. 2009. *Numerical Techniques for Direct and Large-Eddy Simulations.* New York: CRC Press.

Johnston, C. O., P. A. Gnoffo, and A. Mazaheri. 2012. "A study of ablation-flow field coupling relevant to Orion heat shield." *Journal of Thermophysics and Heat Transfer* 26 (2).

Khakpour, Yasmin, and Jamal Seyed-Yagoobi. 2015. "Evaporating liquid film flow in the presense of micro-encapsulated phase change materials:A numerical study." *Journal of Heat Transfer* 021501-1-9.

Kuo, K. K. 2005. *Principles of Combustion.* Hoboken, New Jersey: John Wiley and Sons, Inc.

Lam, K., W. Q. Gong, and R. M. G. So. 2008. "Numerical simulation of cross-flow around four cylinders in an in-line square configuration." *Journal of Fluids and Structures* 24: 34-57.

Law, Chung. 2006. *Combustion Physics.* New York: Cambridge University Press.

Lee, W. H. 1980. "A pressure iteration scheme for two-phase flow modeling." In *Multiphase Transport Fundamentals, Reactor Safety, Applications*, by T. N. Verioglu. Washington, DC: Hemisphere.

Mei, Zongshu, Shi Chenging, Xueling Fan, and Xianbin Wang. 2020. "Numerical simulation of hypersonic reentry flow field with gas-phase and surface chemistry models." *Materials Today Communications* 22: 100773.

Michaelides, E. E. 2006. *Particles, Bubbles, and Drops Their Motion, Heat and Mass Transfer.* Hackensack, NJ: World Scientific Printers.

Nellis, Gregory, and Sanford Klein. 2012. *Heat Transfer.* New York: Cambridge University Press.

Nellis, Gregory, and Sanford Klein. 2012. *Thermodynamics.* New York: Cambridge University Press.

Park, C., and H. K. Ahn. 1999. "Stagnation-point heat transfer rates for Pioneer-Venus probes." *Journal of Thermophysic Heat Transfer* 13(1): 33-41.

Qui, G. D., W. H. Cai, Z. Y. Wu, Y. Yao, and Y. Q. Jiang. 2015. "Numerical simulation of forced convective condensation of propane in a spiral tube." *Journal of Heat Transfer* 041502-1-9.

Ray, R., F. Durst, and V. Kumar. 2007. "Modeling moving-boundary problems of solidification and melting adopting an arbitrary lagrangian-eulerian approach." *Numerical Heat Transfer, Part B: Fundamentals* 299-331: 299-331.

Selvnes, H., Y. Allouche, A. Sevault, and A. Hafner. 2018. "CFD modeling of ice formation adn melting in horizontally cooled and heated plates." *Eurotherm Seminar.* University de Lieida. 112.

Shyy, W. 1994. *Computational modeling for fluid flow and interfacial transport.* Mineola, New York: Dover Publications, Inc.

Stavropoulos, P., and P. Foteinopoulos. 2018. "Modeling of additive manufacturing processes: a review and classification." *Manufacturing Review* 2017-014.

Sun, Dongliang. 2014. "Modeling of the evalopration and condensation phase-change problems with FLUENT." *Numerical Heat Transfer A: Fundamentals* 326-342.

Tabbara, H., and S. Gu. 2012. "Modelling of impingement phenomena for molten metallic droplets with low to high veloicities." *International Journal of Heat and Mass Transfer*, 2081-2086.

Voller, V. R., and C. R. Swaminathan. 1987. "A fixed grid numerical modeling methodology for convective diffusion mushy region phase-change problems." *International Journal of Heat and Mass Transfer* 30: 1709-1720.

Welch, W. J., and J. Wilson. 2000. "A volume of fluid based method for fluid flows with phase change." *Journal of Computational Physics* 160: 662-682.

Zhang, Xiao-Bin, Wei Zhang, and Xue-Jun Zhang. 2012. "Modeling droplet vaporization and combustion with the VOF method at a small Reynolds number." *Journal of Zhejiang University - Science A (Applied Physics and Engineering)* 13(5): 361-374.

Zhluktov, S. V., and T. Abe. 1999. "Viscous shock-layer simulation of airflow past ablating blunt body with carbon surface." *Journal of Thermophysics and Heat Transfer* 13 (1): 50-59.

In: Computational Fluid Dynamics ISBN: 978-1-53619-756-3
Editor: James S. Hutchinson © 2021 Nova Science Publishers, Inc.

Chapter 2

CFD-BASED DESIGN OF NOVEL GREEN FLASH IRONMAKING REACTORS

H. Y. Sohn[*]

Department of Materials Science and Engineering,
University of Utah, Salt Lake City, Utah, US

ABSTRACT

A critical problem facing the steel industry is to develop a new technology to produce iron with significantly reduced energy consumption and greenhouse gas emissions. The development of a novel ironmaking technology based on direct utilization of fine iron ore concentrate in a flash reactor is summarized. The CFD-based design of potential industrial reactors for flash ironmaking is described. Overall, this work has shown that the size of the reactor used in the novel Flash Ironmaking Technology (FIT) can be quite reasonable vis-à-vis the blast furnaces. As an example, a flash reactor of 12 m diameter and 35 m with a single burner operating at atmospheric pressure would produce 1.0 million tons of iron per year. The height can be further reduced by using multiple burners and/or preheating the feed gas.

---

[*] Corresponding Author's Email: h.y.sohn@utah.edu.

The CFD-based design of potential industrial reactors for flash ironmaking pointed to a number of features that an industrial reactor should incorporate. The flow field should be designed in such a way that a larger portion of the reactor is used for the reduction reaction but at the same time excessive collision of particles with the wall must be avoided. Further, a large diameter-to-height ratio should be used from the viewpoint of decreased heat loss. This may require the incorporation of multiple burners and solid feeding ports on the reactor roof.

Keywords: concentrate, energy consumption, flash ironmaking technology (FIT), hydrogen, kinetics, magnetite, natural gas, CFD simulation, reactor design, burner, partial combustion

INTRODUCTION

A critical problem facing the steel industry is the development of an innovative technology for producing iron in the future that is much more energy-efficient and environmentally friendly. This article describes the development of a novel Flash Ironmaking Technology (FIT), conceived by Sohn [1], which is based on the reduction of iron oxide concentrate particles by gaseous fuel/reductant in a vertical flash reactor. The novel process addresses the critical issues in ironmaking, i.e., energy saving and greenhouse-gas emissions. The steel industry is responsible for about 6~7% of total human-made emissions of carbon dioxide [2].

The alternative ironmaking Direct Reduction processes [3] are divided into two different categories: shaft furnace processes (Midrex and Energiron [4]) and fluidized-bed processes (e.g., FIOR [5], FINMET [5], CIRCORED [6], and SPIREX [7]). These processes are not intensive enough to compete with the blast furnace. The shaft furnace processes must use pellets of iron oxide concentrate that are expensive to produce, consume energy, emit pollutants including CO_2, and frequently experience the problems of pellet breakage.

A novel Flash Ironmaking Technology (FIT) has been conceived by Sohn [1] developed for producing iron directly from fine concentrates by a flash reduction process. This process uses a reductant gas such as natural gas or hydrogen and does not require pellets, sinters or coke required in other ironmaking processes [1, 8, 9, 10]. The new technology would reduce energy consumption by 30-60% and decrease carbon dioxide emissions by 60-96% compared with the blast furnace ironmaking, depending on whether hydrogen or hydrocarbon gas is used. The FIT will not have the problems like particle sticking or pellet disintegration. High-grade lump iron ore is scarce world-side, and new reserves must be ground to finer sizes to beneficiate them. Thus, increasing amounts of concentrates that can feed the FIT reactor are expected to be produced world-wide [11].

Based on the potential advantages discussed here and the results of the process feasibility studies, detailed flow sheets for different versions of the new process depending on the fuel type and economic analysis have been developed by Pinegar et al. [12, 13, 14, 15].

DESCRIPTION OF FLASH IRONMAKING TECHNOLOGY (FIT)

A sketch of the new flash ironmaking process is shown in Figure 1. In this process, the fuel gas is partially burned with tonnage oxygen which generates a reducing gas of 1500 – 1800 K. The injected concentrate particles are reduced as they move downward. The process may be operated to create a molten-iron bath for possible direct steelmaking or to produce solid iron particles that can be charged in subsequent steelmaking furnaces.

The Flash Ironmaking Technology (FIT) is expected to eliminate several of the critical issues accompanying other alternate ironmaking processes such as (a) the requirements of pelletization or sintering; (b) cokemaking; (c) solids sticking or disintegration; and (d) refractory erosion. This technology offers the possibility to produce either solid iron particles or molten iron that leads to direct steelmaking in a single unit, as shown in Figure 1.

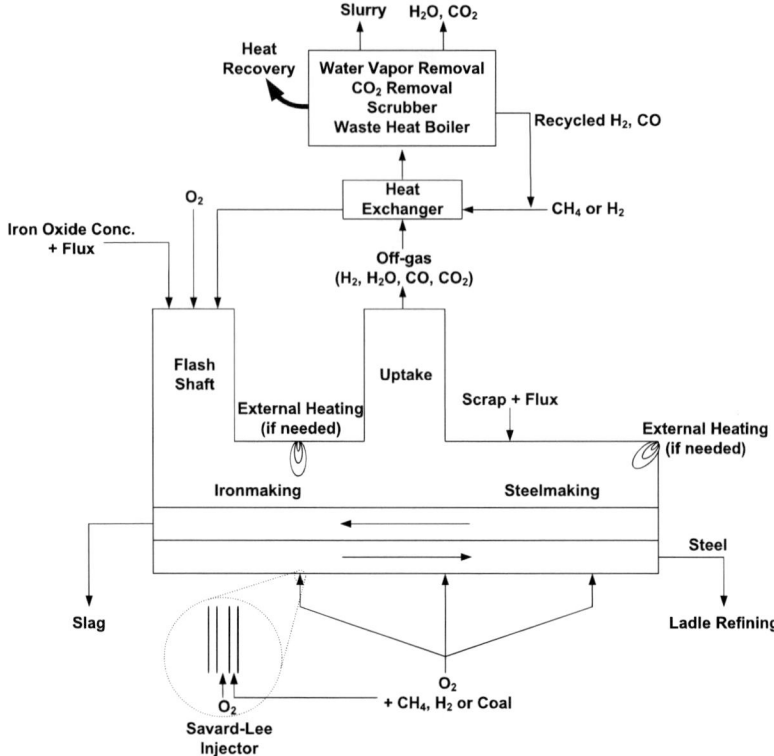

Figure 1. A schematic diagram of a possible direct steelmaking process based on Flash Ironmaking Technology (FIT).

Reduction Kinetics of Magnetite Concentrate Particles

Hydrogen is the main reducing agent in the new process that takes place at temperatures above 1473 K, even when an H_2+CO mixture produced by the partial combustion of natural gas is used. Carbon monoxide is highly stable at these high temperatures and thus its direct contribution is much less from the viewpoints of thermodynamics and also kinetics relative to hydrogen. However, it provides a synergistic effect on the hydrogen reduction, as shown below. Sohn and coworkers [16, 17, 18, 19, 20, 21] have investigated the reduction rates of magnetite and hematite concentrates under the temperature ranges and gaseous reactant partial pressures applicable to the Flash Ironmaking Technology (FIT), using laminar-flow reactors. The results were found to yield rate equations that can be expressed by the following general form of $\frac{dX}{dt}$ for both component gases H_2 and CO:

$$\left.\frac{dX}{dt}\right|_j = k_j \cdot \left[p_j^{m_j} - \left(\frac{p_{jo}}{K_j}\right)^{m_j}\right] \cdot n_j (1-X)[-\text{Ln}(1-X)]^{1-1/n_j} \cdot d_p^{-s_j}; j$$
$$= H_2 \text{ or CO} \tag{1}$$

where k_j is the reaction rate constant for gas j, $k_j = k_{o,j} \exp\left(-\frac{E_j}{RT}\right)$; p_j is the partial pressure of gas j; K_j is the equilibrium constant for the reduction of FeO by gas j; m_j is the reaction order with respect to gas j; n_j is the Avrami parameter; $d_p^{-s_j}$ is the particle size effect function; X is the reduction degree defined as the ratio of the removed oxygen to the total removable oxygen in the concentrate particles.

The rate parameters that are most appropriate for application in designing and analyzing the operation of a flash ironmaking reactor are given below. The reader is referred to the original papers for other details of the rate measurement and data analysis.

Table 1 lists the kinetic parameters for reduction of magnetite concentrate by individual component gases.

Table 1. Kinetic parameters for reduction of magnetite concentrate by individual component gases [16, 17, 20, 21]

Reducing Gas, j	Temperature Range	$k_{o,j}$	E_j (kJ/mol)	m_j	n_j	s_j
H_2	1423 - 1623 K	1.23×10^7 atm^{-1} s^{-1}	196	1	1	0
	1623 - 1873 K	6.07×10^7 atm^{-1}·s^{-1}·μm	180	1	1	1
CO	1423 - 1623 K	1.07×10^{14} atm^{-1} s^{-1}	451	1	0.5	0
	1623 - 1873 K	6.45×10^3 atm^{-1}·s^{-1}·μm	88	1	0.5	1

When magnetite concentrate is reduced by a mixture of H_2 + CO, the CO enhances the rate of reaction between H_2 and iron oxide, most likely due to the effect of CO on the morphology of the reduced iron by forming whiskers which was observed in a separate study [22]. Taking this into consideration, Fan et al., [17] developed the following rate expression:

The complete rate equations for the reduction of magnetite concentrate particles by H_2 + CO mixtures in the temperature ranges 1423 K (1150°C) - 1623 K (1350°C) and 1623 K (1350°C) - 1873 K (1600°C) are given, respectively, as:

$$\frac{dX}{dt} = \left(1 + 1.3 \cdot \frac{p_{co}}{p_{co}+p_{H_2}}\right) \cdot \frac{dX}{dt}\bigg|_{H_2} + \frac{dX}{dt}\bigg|_{CO} \quad 1423\ K < T < 1623K \quad (2)$$

$$\frac{dX}{dt} = \left[1 + (-0.01T + 19.65) \cdot \frac{p_{co}}{p_{co}+p_{H_2}}\right] \cdot \frac{dX}{dt}\bigg|_{H_2} + \frac{dX}{dt}\bigg|_{CO} \quad 1623\ K < T < 1873K \quad (3)$$

where $\frac{dX}{dt}\big|_{H_2}$ and $\frac{dX}{dt}\big|_{CO}$ represent the rates of reduction individually by H_2 and CO, respectively, obtained from Eq. (1) with the parameters listed in Table 1.

Similar rate measurements were also done with hematite concentrate [18, 23, 24]. It was conclusively confirmed by the work described above that iron ore concentrate particles can be > 95% reduced by hydrogen in several seconds of residence time typically available in a flash reactor at 1473 K or above.

Tests in a Laboratory Flash Reactor

Based on the results of kinetic feasibility discussed above, reduction tests were conducted by Sohn et al., [8, 25, 26] in a laboratory flash reactor shown in Figure 2. An industrial flash reactor would be significantly different from a laminar-flow reactor (LFR) used for the rate measurement, including the fact that an oxy-fuel burner would be the main source of heat and the amount of excess reducing gases would be much lower (20-100%). The laboratory flash reactor had many of the features of an industrial flash reactor. Other experimental details can be found elsewhere [25, 26, 27, 28].

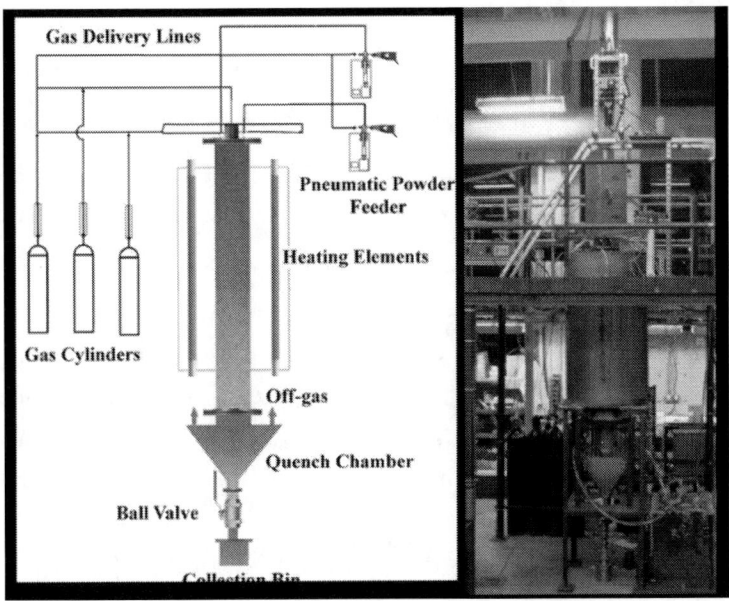

Figure 2. The Utah laboratory flash ironmaking reactor (I.D. 0.19 m and length 2.13 m).

Experiments were performed with either hydrogen or methane and with different modes of gas and particle feeding, the results of which were analyzed with the aid of computational fluid dynamics (CFD) simulations.

The tests and simulation using the laboratory flash furnace produced the useful information, the most important of which can be summarized as follows:

- Iron can be obtained from iron concentrates by flash reduction using partial combustion of gaseous fuels hydrogen, natural gas or a mixture thereof,

- The configuration of fuel gas and oxygen feeding is important in terms of temperature uniformity and accompanying concentrate feeding mode. A flame generated by central feeding of the gaseous fuel surrounded by oxygen flow promoted temperature uniformity and allowed solid feeding through the center of the flame.
- The best position for concentrate feeding is near but outside the flame. This configuration works with either mode of gas feeding, fuel gas surrounded by oxygen or vice versa.

Operation of Pilot Plant with Flash Reactor

A Pilot Flash Reactor (PFR) capable of operating at 1200-1600°C with a concentrate feeding rate of 1-7 kg/h, shown in Figure 3, was operated at the University of Utah [27]. This reactor was the first flash ironmaking reactor where the heat and reductant are produced by partial oxidation of natural gas or hydrogen with tonnage oxygen.

Figure 3. The pilot plant with a Flash Reactor installed at the University of Utah.

CFD-Based Design of Novel Green Flash Ironmaking Reactors 63

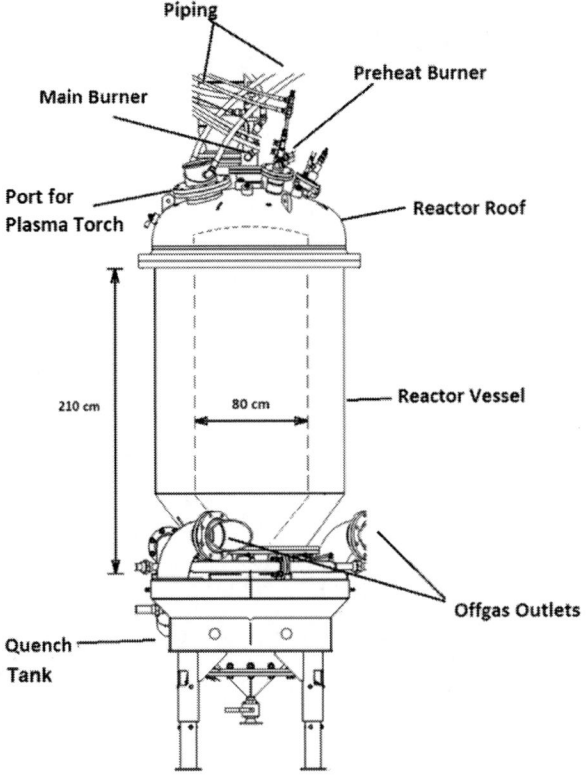

Figure 4. Schematic diagram of the Pilot Flash Reactor.

The PFR consisted of a reactor vessel, a vessel roof, burners, a quench tank, off-gas piping, a flare stack, an off-gas analyzer, a gas valve train, a water cooling system, gas leak detectors, a concentrate feeding system, and human machine interface. Figure 4 shows the main components of the reactor body. Only the very salient features will be described here, and other details of the facility and its operation are described elsewhere [27].

Burners

The PFR had 3 burners: a preheat burner, the main burner, and a plasma burner. Figure 5 shows a schematic diagram for the cross section of the preheat burner and the main burner.

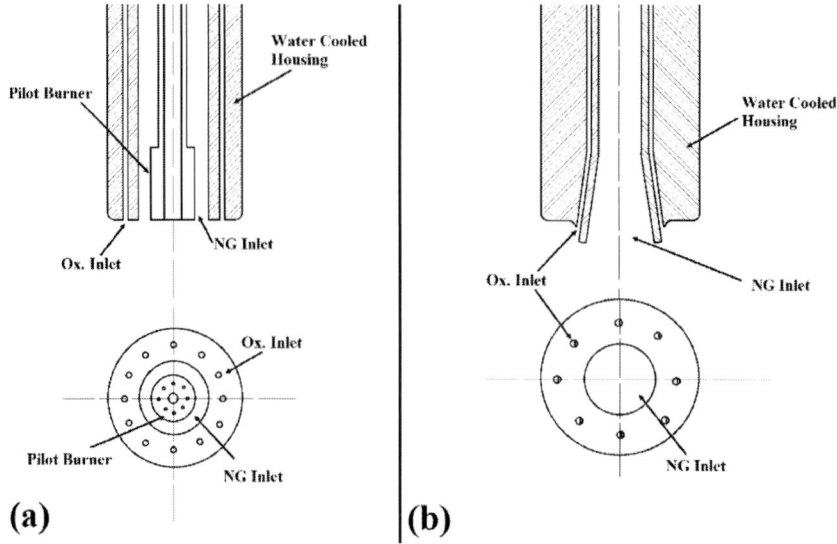

Figure 5. Schematic diagrams for (a) preheat burner and (b) main burner.

Concentrate Feeding System

The concentrate was fed into the reactor using a HA5171P-D powder feeder supplied by HAI, Placentia, CA, U.S. This pneumatic powder feeder was used to feed magnetite concentrate to the reactor at a feeding rate of 1-7 kg/h. Nitrogen gas at a flow rate of 11 SLPM was used as the carrier gas. The particles were fed through feeding inlets on the sides of the main burner. This was determined based on the results obtained from the laboratory flash reactor.

Human Machine Interface

The Human Machine Interface (HMI) consists of the main PLC and a computer. The main PLC was connected to all the different parts of the system and to the computer where the operator could monitor all the different parts and run the reactor. The programming in the main PLC was responsible for all the safety and emergency steps. The main PLC was supplied by ACS company, Boise, ID, U.S. Figure 6 shows a screen shot where all the parameters of the reactor were displayed and controlled.

CFD-Based Design of Novel Green Flash Ironmaking Reactors   65

Operation of the PFR

All the components of the reactor were installed and a leak test was performed on the vessel by capping the off-gas pipes and pressurizing the system to 2.0 atm for 45 minutes to make sure that there were no leaks from any components. The system was preheated to the target temperature with a heating rate of 90-95°C/h which was the maximum heating rate that could be used to avoid any damage to the refractories. The heating cycle was automatically controlled by the HMI and the flow rates of natural gas and oxygen were varied based on the measured temperature from the reactor vessel.

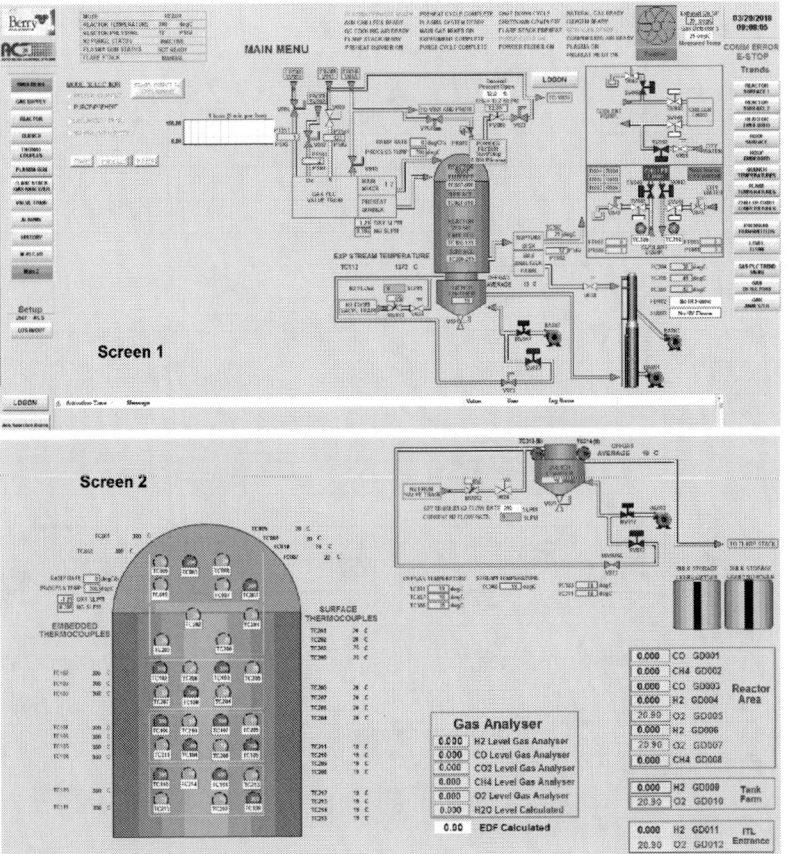

Figure 6. The controlling screen for the HMI.

Results from PFR Runs

This reactor was simulated by CFD to optimize the operating conditions to achieve high reduction degree at the optimum conditions and reactor sizes to be used in an industrial reactor [29]. The results obtained from the developed CFD model was compared to the actual results of the reactor operation and good agreement was achieved. Table 2 shows the results of the runs performed in the reactor.

Table 2. The results of runs performed in the PFR

Inner Wall Temperature (°C)	Magnetite Concentrate Feeding Rate (kg/h)	Gas Flow Rate (SLPM)* Main Burner		H_2 EDF	Nominal Residence Time (s)	RD (%)
		NG (SLPM)	O_2 (SLPM)			
1200-1130	5.0	404	321	0.76	12.5	65
1290-1220	1.8	410	293	0.84	12.0	79
1290-1210	2.9	410	293	0.96	12.0	82
1290-1230	2.5	358	270	1.00	13.3	83
1290-1240	3.5	512	327	1.07	10.2	76
1330-1230	4.7	330	200	1.36	15.3	89
1330-1230	4.5	330	200	1.44	15.3	87
1330-1230	5.2	500	290	3.00	10.6	80
1330-1230	4.3	500	290	3.00	10.6	82
1355-1260	5.5	235	190	0.03	18.3	7
1350-1300	4.0	255	209	0.15	17.0	49
1350-1270	4.5	275	212	0.20	16.2	31
1340-1280	5.0	280	209	0.21	16.2	37
1350-1290	4.6	280	230	0.50	15.6	80
1400-1300	6.3	300	240	0.82	14.4	88
1400-1300	5.0	330	200	1.51	14.6	100
1415-1350	4.5	220	191	0.07	18.0	18
1410-1360	4.0	240	195	0.33	17.1	32
1410-1330	5.0	295	221	0.50	14.7	66
1410-1330	6.0	300	210	0.70	14.9	74
1410-1320	5.0	300	210	0.82	14.9	82

*The flow rates of NG and O_2 in the pilot burner were 9.6 and 37.6 SLPM, respectively. The flow rate of N_2 in the powder feeder was 10.7 SLPM.

Different experimental runs were designed in this reactor to yield a wide range of reduction degree at less than complete reduction to better examine the effects of the operating conditions and validate the CFD model in these different conditions. The results showed good reproducibility within ±5% of the average reduction degree. This represents a very high degree of reproducibility, considering the complexity of the operation and design of this large unit.

DESIGN OF MEDIUM-SIZE FLASH IRONMAKING REACTORS

A reactor with a capacity of 100,000 tons/yr of iron was designed to further test the feasibility of FIT in this range of production rate [28]. For the proper design and scale-up of such reactors, it is essential to have information on the temperature and species distribution, gas and particle flow patterns. This information is difficult or even impossible to obtain from experiments. With computational fluid dynamics (CFD) modeling, it is possible to gain such insights on these critical parameters that are essential in reactor design.

Here, two types of reactors were designed. The first type was to produce metallic iron in solid state. The typical operating temperature in this case is around 1300°C. The solid-state product collected could be charged into an electric arc furnace in the steelmaking process. The second type is to produce iron in molten state, which is typically operated at a temperature of around 1600°C, and can lead to direct steelmaking combined with flash reduction or charged into a basic oxygen furnace or an electric arc furnace without further treatment.

Geometries and Dimensions

Sketches of possible configurations of flash ironmaking reactors are shown in Figure 7. Depending on the operating conditions, the main body of the reactor is either made up of a cylindrical part and a conical part or a cylindrical shaft only. Under solid state operating conditions, a conical part near the exit of the reactor is needed for solid particle collection. If iron is produced in molten state, a bath settler is needed below the shaft, as shown in Figure 1. Our focus is mainly on the shaft part of the reactor in this work as the reduction of concentrate particles mostly happens during their travel in the shaft.

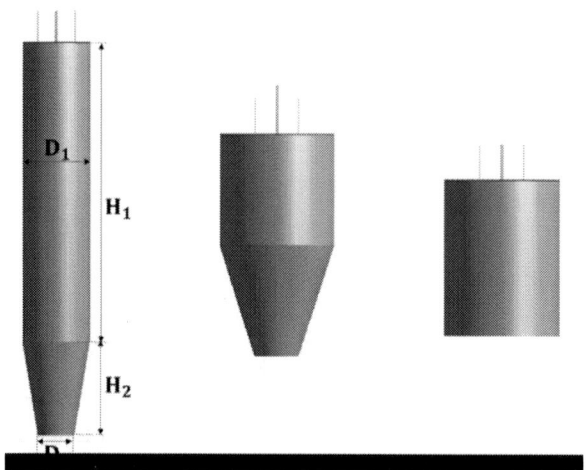

Figure 7. Sketches of possible configurations of flash ironmaking reactors.

Table 3. Dimension of reactors with one-burner

D_1 (m)	D_2 (m)	H_1 (m)	H_2 (m)	Preheat Temp. (°C)	Designed Product Temp. (°C)	Design #
4.0	2.0	12.0	6.0	600	1300	1
4.0	2.0	10.0	6.0	1000	1300	2
6.0	2.0	6.0	6.3	600	1300	3
6.0	2.0	6.0	5.0	1000	1300	4
4.0	--	13.0	--	1000	1600	5
6.0	--	9.0	--	1000	1600	6

With the same reactor volume, the design with a large height to diameter ratio leads to a long and thin reactor, while a small height to diameter ratio leads to a short and fat one as shown in Figure 7. In this study, two typical diameters, 4 m and 6 m, were tested. The diameter of the long and thin reactor was set to be 4 m. A diameter of 6 m was used for the short and fat reactor. The number of burners to be used is also an important factor in reactor design. Reactors with one burner and four burners were tested in this work. Before deciding the number of burners to be used, the optimal value for the diameter was first determined under the one-burner design. The dimensions of the reactors simulated are listed in Table 3. The powders were fed through four feeding ports installed on the roof of the reactor. The four powder feeding ports were distributed evenly (90 degrees apart), as shown in Figure 8. The distance between each feeding port and the centerline of the reactor was equal to half of the radius.

A nonpremixed burner with two oxygen slots and one natural gas slot, shown in Figure 9, was used in the simulation. The reactor wall consisted of three layers, namely, the refractory layer, insulation layer and steel shell layer from the inner layer to the outer layer, as shown in Figure 10. The thickness of the refractory layer, insulation layer and steel shell layer are kept at 0.15 m, 0.08 m and 0.0254 m, respectively. Wall materials at those thicknesses were proved to be efficient in a large bench flash ironmaking reactors constructed on the campus of University of Utah that was designed to operate from 1200°C to 1600°C. The properties of the wall materials are listed in Table 4.

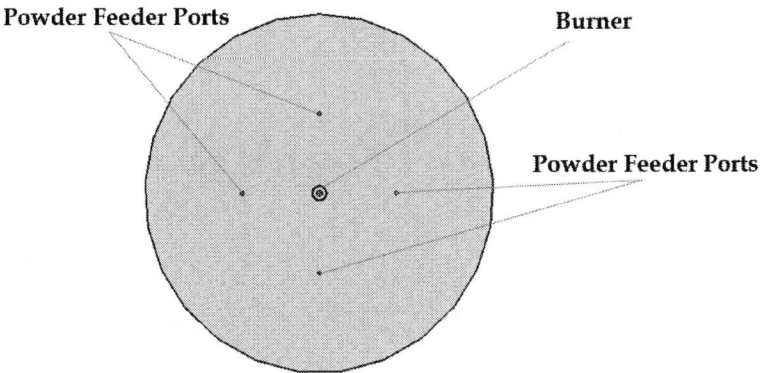

Figure 8. Distribution of the powder feeding ports on the roof of the reactor.

Table 4. Wall material properties

	Thermal Conductivity (W·m⁻¹·K⁻¹)	Density (kg·m⁻³)	Specific Heat (J·kg⁻¹·K⁻¹)
Refractory	$10^{-6} T^2 - 0.0032 T + 4.5396$	2890	$0.2965 T + 362$
Insulation	$3 \times 10^{-8} T^2 + 4 \times 10^{-5} T + 0.1797$	1081	714
Steel shell	50	7850	470

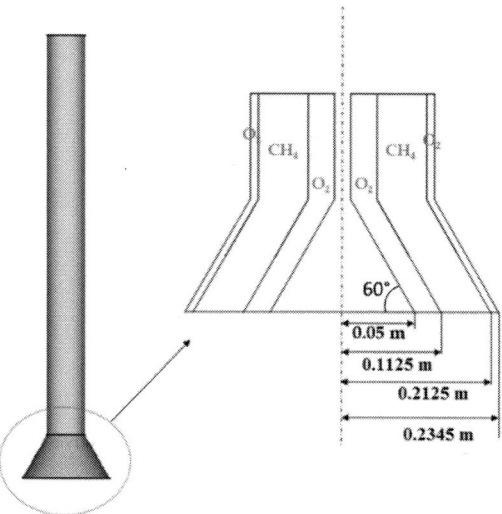

Figure 9. Burner configuration.

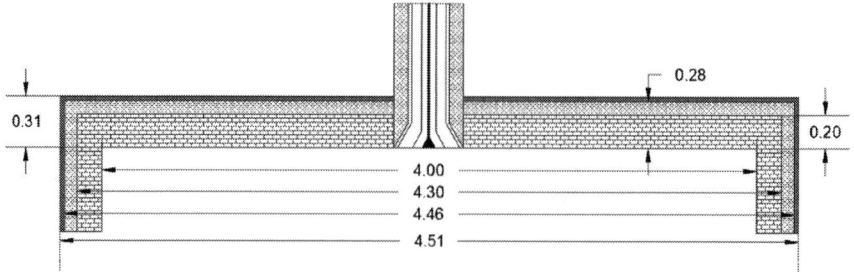

Figure 10. Reactor wall structure (unit in m).

Operating Conditions

The fuel entering the reactor consisted of fresh natural gas and recycled H_2, which were partially oxidized by oxygen to generate the heat needed for the reaction as well as the reducing gases CO and H_2. The actual composition of natural gas was 96% CH_4, 2% C_2H_6, and 2% nitrogen by volume. Natural gas was considered as 98.1% CH_4 (1 mol% of C_2H_6 equivalent to 2.6 mol% of CH_4 in heat production and 2 mol% of CH_4 in hydrogen and carbon monoxide production, both for generating a representative hot gas mixture from the partial combustion. Thus considering these two factors, 1 mol% of C_2H_6 was treated as being equivalent to 2.3 mol% of CH_4) and 1.9% N_2 to avoid the complexity of including the small amount of C_2H_6 in the combustion calculations. The input gases were preheated to a specified temperature before charging into the reactor to reduce the amount of input gases needed. In this work, two preheat temperatures (600°C and 1000°C) were investigated. The operating conditions are summarized in the following Tables 5-7.

The concentrate feeding rate was calculated based on 340 normal operating days in a year, 70 wt.% total iron content in the concentrate and product metallization of 95%.

Steady state conditions were simulated in this work. The Euler-Lagrange approach was used to model the two-phase flow, in which the gas phase was treated as a continuum in the Eulerian frame of reference while the solid phase was tracked in the Lagrangian mode. A two-way coupling approach between the gas phase and solid phase was used in the simulation. The CH_4-O_2 combustion mechanism available in the literature [30] was used. In a large reactor, efficient cooling method, such as copper stove cooling technology, is usually necessary to cool the outer surface of the reactor to an acceptable temperature. The incorporation of such cooling system significant complicated the model. For simplicity, the outer surface of reactor in this work was set to be 60°C for all the calculations.

Table 5. Operating conditions for solid product with input gases preheated to 600°C

	Flow Rate (kg/s)	Preheat Temp. (°C)
Natural Gas	1.15	600
Recycled H_2	0.43	600
Oxygen	2.19	600
N_2 (Carry Gas)	0.07	25
Concentrate	5.20	25

Table 6. Operating conditions for solid product with input gases preheated to 1000°C

	Flow Rate (kg/s)	Preheat Temp. (°C)
Natural Gas	0.91	1000
Recycled H_2	0.36	1000
Oxygen	1.54	1000
N_2 (Carry Gas)	0.07	25
Concentrate	5.20	25

Table 7. Operating conditions for molten product with input gases preheated to 1000°C

	Flow Rate (kg/s)	Preheat Temp. (°C)
Natural Gas	0.91	1000
Recycled H_2	0.36	1000
Oxygen	1.54	1000
N_2 (Carrier Gas)	0.07	25
Concentrate	5.20	25

RESULTS AND DISCUSSION

The mass weighted average metallization at the exit of reactor in all the designs listed in Table 3 were aimed to be 95%.

One-Burner Design

Typical velocity fields in the plane that passes through the center of two powder feeding ports are shown in Figures 11. It is seen from these figures that due to the high-velocity jets erupting from the burner nozzles, recirculation zones formed in regions close to the reactor inner wall in the top part of the reactor. In design # 1, the particles entering the reactor may be pushed to the reactor wall by the hot, high-velocity gas coming out of the flame region as seen from Figures 11 (a). The particles being pushed towards the wall may cause the sticking problem. In design # 3, the concentrate particles were less affected due to its larger diameter. Design # 3 in this case gave a better flow field.

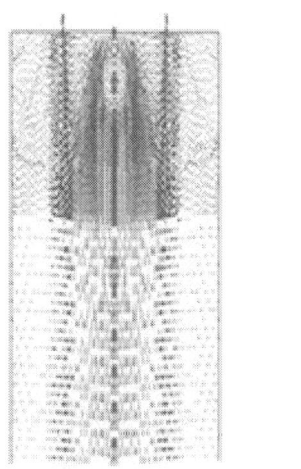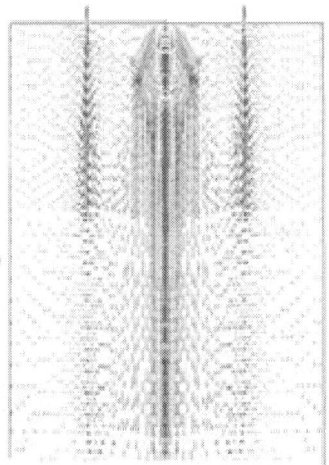

Figure 11. Velocity field in the plane passing through the centers of two powder feeding ports: (a, left) design # 1, (b, right) design # 3 under operating conditions listed in Table 5.

Heat loss is another criterion to look at in the reactor design. The heat loss through the walls in each design was calculated and listed in Table 8. In order to help better evaluate the energy efficiency of each reactor, the percentage of the energy loss (percentage of heat loss from the heat generated from combustion plus amount of sensible energy of the input gases) is also calculated. The numbers indicated that reactors with a diameter of 6 m had a smaller value of heat loss than reactors with a diameter of 4 m, as expected, but this result gives a numerical indication of how the heat loss compares between the two cases.

Therefore, the geometry used in design # 6 is used as the design of the flash reactor that is operated to produce molten iron. The design of a flash reactor that operates to produce solid product will be further discussed in the next section. The species distributions in designs # 5 and # 6 are shown in Figures 12 and 13. The main component gases outside the flame region reached equilibrium quickly and were uniformly distributed. The mole fractions of H_2, H_2O and CO at the exit of two reactors (outside the flame region) were the same at 0.48, 0.32 and 0.13, respectively.

Table 8. Heat loss and heat generated from partial combustion

Design #	Heat Loss (MW)	Heat generated (MW)	Sensible heat of input gases (MW)	Percentage (%)
1	0.72	19.26	6.82	2.76
2	0.55	12.31	9.92	2.47
3	0.62	19.26	6.82	2.38
4	0.45	12.31	9.92	2.02
5	0.68	20.58	13.6	1.99
6	0.61	20.58	13.6	1.79

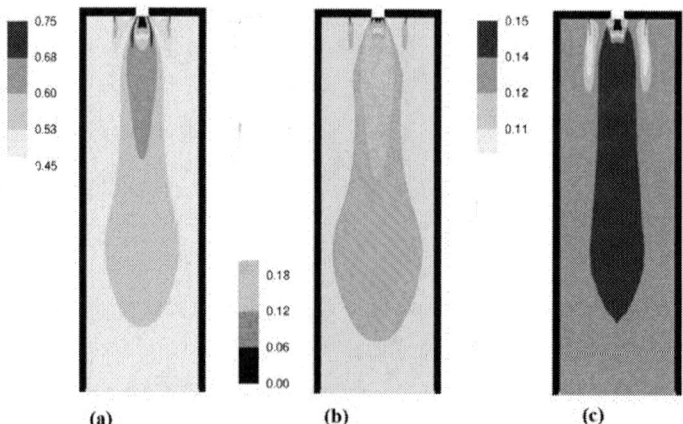

Figure 12. Species distribution in the plane passing through the centers of two powder feeding ports of design # 5: (a) H_2, (b) H_2O, (c) CO.

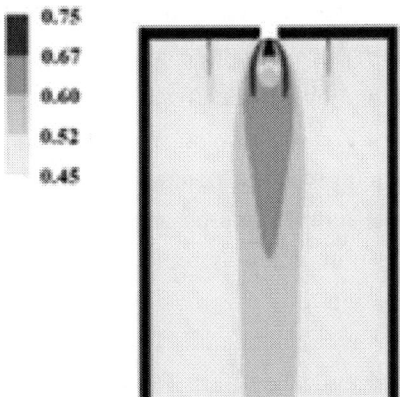

Figure 13. H_2 distribution in the plane passing through the centers of two powder feeding ports of design # 6.

Four-Burner Design

The flash ironmaking reactor with the one-burner had uneven distribution of gaseous species and temperature as well as high particle concentrations near the wall in the top part of the reactor due to the strong recirculation flow. In this section, the four-burner design is discussed. The distribution of the four burners on the roof of the reactor is shown in Figure 14.

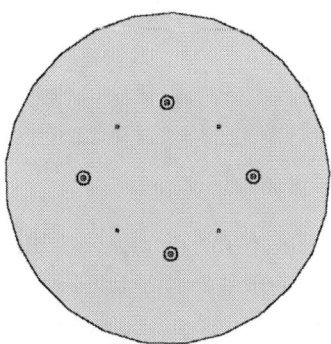

Figure 14. Distribution of the burners on the roof of the reactor. (The large opening are burners and the small openings are powder feeding ports).

The four burners were evenly distributed on the roof 90 degrees apart. The distance between the burner and the centerline of the reactor was equal to half of the radius. The powder feeding ports were symmetrically placed in-between two burners. The distance between the powder feeding port and the centerline of the reactor was also equal to half of the radius. The burners used in this case were different from the one used in the one-burner design. The radial velocity was eliminated by replacing the conical burner tip with a straight concentric design as shown in Figure 15. The natural gas stream was in the middle and was surrounded by two oxygen streams.

The same dimension as design # 3 in Table 3 was used in this simulation. The reactor diameter was chosen as 6 m. The same three layers of walls were also used in this design. The operating conditions listed in Table 5 were used as the reactor was designed to produce solid iron particles.

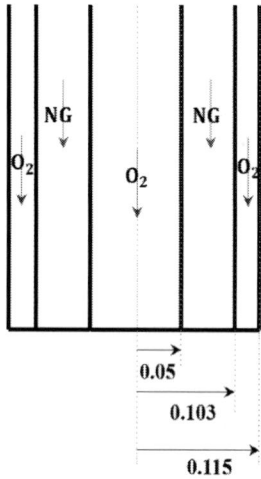

Figure 15. Burner configuration and dimension.

The velocity field in the plane that crosses the center of two powder feeding ports is shown in Figure 16. The radial velocity component near the burner was smaller than that in the one-burner design. The particle number density distribution is shown in Figure 17. The number density of the concentrate particles close to the wall was greatly reduced, making the particle sticking to the wall much less likely.

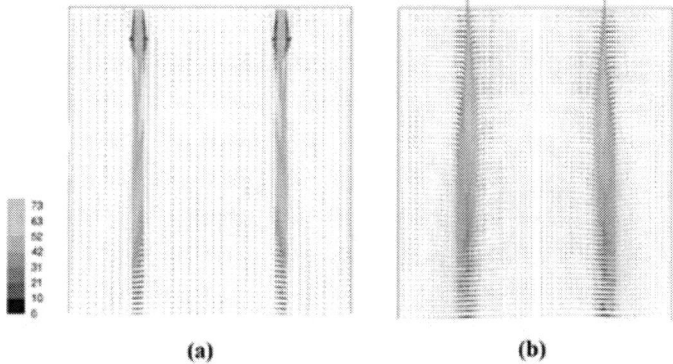

Figure 16. Velocity field in the plane passing through (a) the centers of two opposite burners, (b) the centers of two opposite powder feeder ports (unit in m/s).

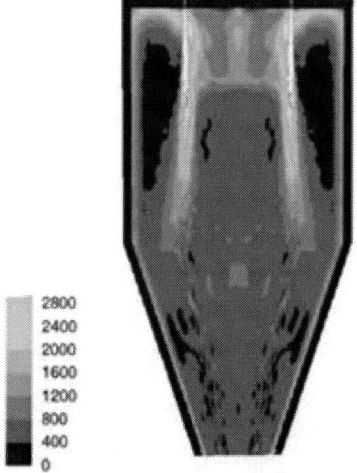

Figure 17. Particle number density (particles/cm^3) in the plane passing through the centers of two opposite powder feeding ports.

The temperature distributions indicated that the particle stream regions in the four-burner design were less exposed to the high temperature of the flame than with the one-burner design,. The consequence of this is that melting of the concentrate particles was less likely to happen so that the reduction of the concentration particles was not affected. Better temperature homogeneity was also seen in this reactor as four burners were used. As a result, the energy generated from the partial combustion was more uniformly distributed inside the reactor. The averaged product temperature at the exit of the reactor was 1278°C rendering a mass averaged reduction degree of the product to be 93%. The heat loss of this reactor was 0.65 MW.

The main species distribution inside the reactor is shown in Figures 18 and 19. No noticeable change in the CO and CO_2 mole fractions outside the flame region was seen as the reduction of magnetite concentrate particles was done by H_2. The mole fractions at the exit of the reactor for H_2, H_2O, CO and CO_2 were 0.47, 0.31, 0.13 and 0.025, respectively.

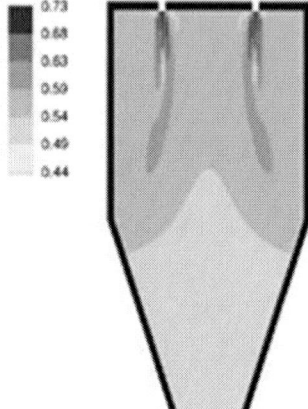

Figure 18. H_2 distribution in the plane passing through the centers of two opposite burners.

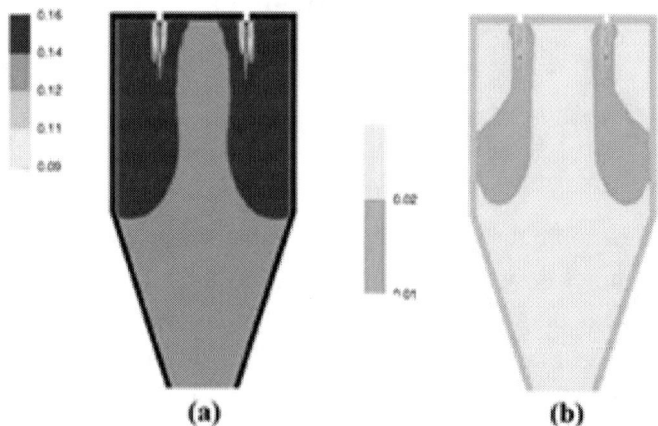

Figure 19. Species distribution in the plane passing through the centers of two opposite burners: (a) CO, (b) CO_2.

Summary

Flash ironmaking reactors of different geometrical dimensions with a capacity of producing 100,000 tons/yr of metallic iron were designed. The metallization degrees of product from these reactors were sufficiently high for use in the subsequent steelmaking step. In the one-burner design, reactors with a diameter of 6 m gave better particle and temperature distributions than reactors with a diameter of 4 m. Better energy efficiency in terms of heat loss was also seen for reactors with a diameter of 6 m. A high particle number density near the wall was less likely in design # 6. A reactor with a diameter of 6 m and 4 burners was also simulated. The larger burner number led to a better particle distribution. The particle distribution in this reactor showed a lower probability of particle sticking compared with design # 3 with a single burner design. All these results may be expected qualitatively, but the CFD simulations present the possibility of yielding quantitative effects of design variation.

DESIGN OF INDUSTRIAL FLASH IRONMAKING REACTORS

An industrial ironmaking plant should have a capacity to produce at least 0.3-1.0 million tons/yr of iron to be competitive with the modern blast furnaces, which typically produce 0.3–3.0 million tons/yr of iron. The same model that was previously described was used for designing two industrial reactors capable of production in this range [28]. A multi-burner configuration was shown in the previous section to have a number of advantages, but for the larger reactors the computational time and difficulties were rather prohibitive for this work. Thus, the industrial reactors were designed to have a single burner in the center of the reactor with four feeding ports. Further, the simulated reactor was designed to produce solid iron particles as opposed to molten iron, as the near-term application of the novel Flash Ironmaking Technology is to produce DRI rather than to operate for direct steelmaking as shown in Figure 1. It is hoped that what we learn from this work will provide helpful information and insight in designing future industrial flash ironmaking reactors.

Dimensions and Operating Conditions

Table 9 shows the operating conditions of the two industrial reactors. The smaller reactor produces 0.3 million tons of iron per year while the larger reactor produces 1.0 million tons of iron per year.

Schematic representations of the reactor and the burner are shown in Figures 20 and 21, respectively, while the dimensions for the two reactors are shown in Table 10.

Table 9. Operating conditions of the two industrial reactors

Parameter	Reactor 1	Reactor 2
Target production of iron in million tons/yr.	0.3	1.0
Feed of the magnetite concentrate in million tons/yr.	0.415	1.38
Natural gas feeding rate in m^3/s	17.5	45.8
Oxygen feeding rate in m^3/s	13.2	34.7
Expected excess of hydrogen at full reduction (EDF)	0.3	0.3

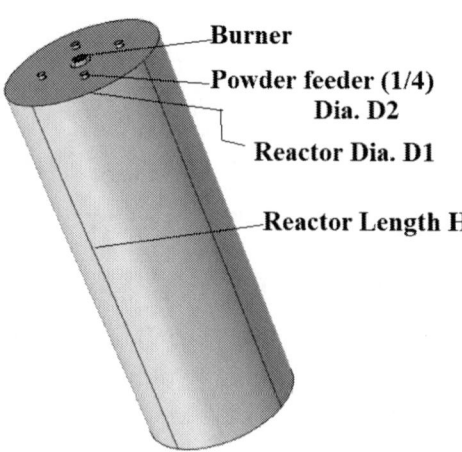

Figure 20. Schematic representation of the industrial reactor.

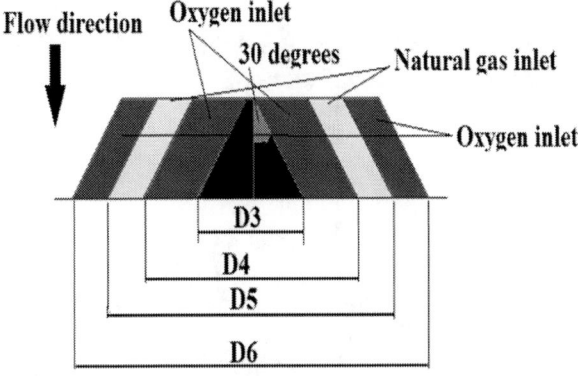

Figure 21. Schematic representation of the burner.

Table 10. The dimensions of the industrial reactors

Parameter	Definition	Reactor 1 (m)	Reactor 2 (m)
H	Height	35.00	35.00
D1	Inner diameter	7.00	12.00
D2	Diameter of powder feeder (4 feeders)	0.05	0.30
D3	Inner diameter of the oxygen inlet 1	0.02	0.02
D4	Outer diameter of the oxygen inlet 1	0.26	0.8
D5	Outer diameter of the natural gas inlet	0.44	1.6
D4	Outer diameter of the oxygen inlet 2	0.51	1.8

It is noted that the radial location of the powder feeders was half the radius of the reactor. Also, the volumetric flow rates of oxygen in inlets 1 and 2 were equal.

Meshing

The industrial reactors were designed to have a single burner in the center of the reactor with four feeding ports evenly distributed and have the same radial position equal to half the radius of the reactor. The symmetry of the reactor was used to decrease the computational time by taking a quarter of the reactor as a representation for the entire reactor.

The mesh consisted of 264,000 hexahedral cells in the smaller reactor and 279,000 hexahedral cells in the larger reactor. The top section of the meshing for Reactor 2 is shown in Figure 22.

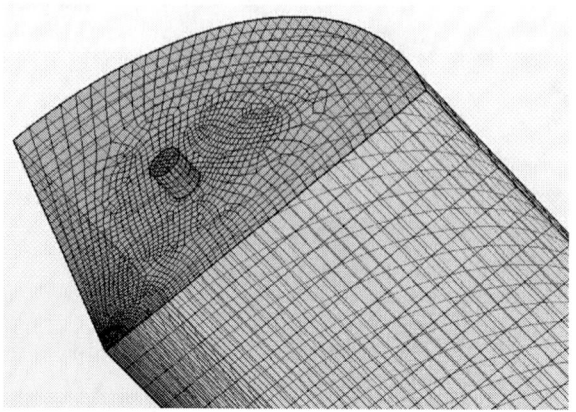

Figure 22. Meshing of the top section for a quarter of Reactor 2.

Results and Discussion

Mass weighted average gas composition and product metallization at the outlet: A velocity of 100 m/s was used for the inlet gases in Reactor 1, while for the larger Reactor 2 the area of the burner was increased and thus the inlet velocity was 37 m/s. The products from reactors at exit can tell us the efficiency of the design. As shown in Table 11, the metallization degrees of the products from the two reactors were nearly identical at > 90%. The higher temperature of the gas mixture in Reactor 2 indicates that heat loss was lower from it, as expected.

Table 11. Average gas composition and metallization degree at reactor exit

Reactor	T (K)	H_2	CO	CO_2	H_2O	Metallization (%)
1	1519	40.2	26.3	5.9	24.1	91.2
2	1578	39.9	26.6	5.7	24.8	91.4

Contours of temperature: Figure 23 shows the temperature distribution.

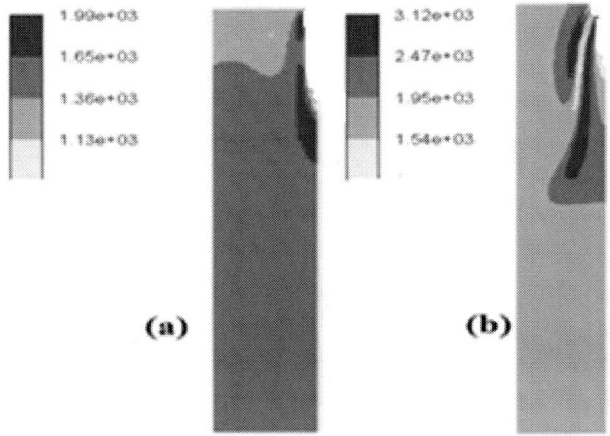

Figure 23. Contours of gas temperature (in K) where the right vertical line represents the axis of symmetry: (a) Reactor 1, (b) Reactor 2.

The temperature distributions in Reactor 1 in Figures 23(a) show a single flame where partial oxidation of natural gas produces the reducing gases. The gas mixture expands because of the increase in the temperature and the total molar flow rate of gas products. Figure 23(b) shows a split flame which is different from the single flame of Reactor 1 in Figure 23(a). The split flame in Reactor 2 arose from the larger thickness of the natural gas stream compared to Reactor 1.

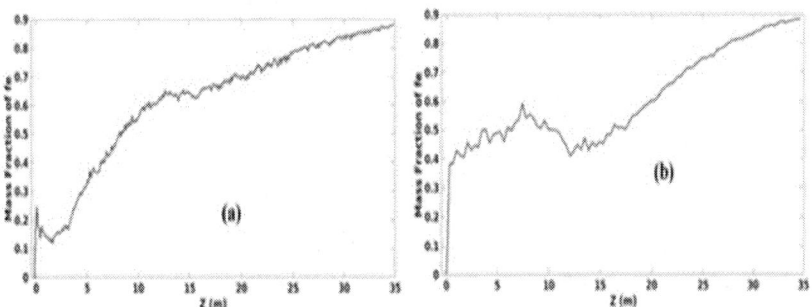

Figure 24. Mass weighted average mass fraction of metallic iron in particles: (a) Reactor 1, (b) Reactor 2.

The mass weighted average of mass fraction of metallic iron in particles and the particle temperature profiles are plotted in Figure 24 and 25, respectively, against the axial distance in the reactors starting at inlet and ending at the outlet of the 35-meter long reactor

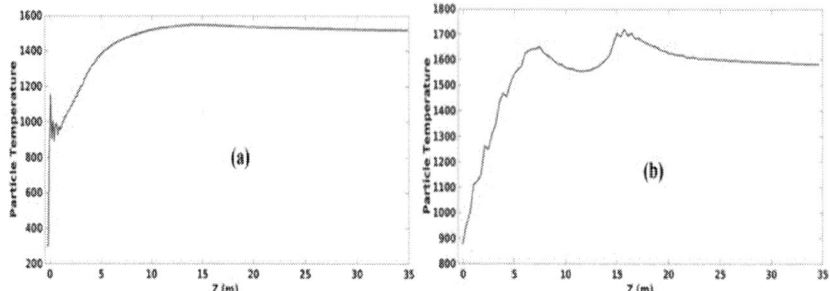

Figure 25. Mass weighted average particle temperature (in K): (a) Reactor 1, (b) Reactor 2.

Table 12. Heat generated from the combustion of natural gas, heat loss from the walls, and percentage heat loss

Reactor	Heat generated (MW)	Heat loss (MW)	Percent heat loss
1	89.8	1.8	2.0
2	327	3.8	1.2

The irregular variations in the curves in Figures 24 and 25 arose from the recirculating flows in the reactors as particles will stay longer on average in those areas. The particle temperature did not reach 1811 K, which is the melting temperature of iron, as we designed the reactor to produce solid iron particles.

Heat loss: The amount of heat generation and the percentage of heat loss in each of the two reactors are summarized in Table 12.

The numbers indicate that the design of Reactor 2 with a smaller surface area per volume lost only about half of the % heat loss of Reactor 1.

Summary

Two industrial reactors with different production rates were designed. The metallization degrees of product from these reactors were sufficiently high for use in the subsequent steelmaking step. The modification in Reactor 2 caused higher outlet gas temperature, more uniform temperature along the reactor. Particle distribution showed that Reactor 2 had a better distribution with a lower likelihood of particle sticking on the wall. Reactor 2 also showed a lower percentage of heat loss compared to Reactor 1 because of the lower surface area per volume.

CONCLUSION

Overall, this simulation work has shown that the size of the reactor used in the novel Flash Ironmaking Technology (FIT), even at the production rate comparable to the largest blast furnaces currently used in the steel industry, can be quite reasonable vis-à-vis the blast furnaces. As an example, a flash reactor of 12 m diameter and 35 m with a single burner operating at atmospheric pressure would produce 1.0 million tons of iron per year. The height can be further reduced by using multiple burners or preheating the feed gas. Further, the total volume of the reactor can be greatly reduced by operating the reactor under elevated pressures, from the points of residence time and reaction kinetics. Obviously, the cost of the reactor per unit volume and those of operation and safety measures would increase accordingly. Thus, the actual design will require optimization by taking into consideration these various factors.

The CFD-based design of potential industrial reactors for flash ironmaking pointed to a number of features that an industrial reactor should incorporate. The flow field should be designed in such a way that a larger portion of the reactor is used for the reduction reaction but at the same time excessive collision of particles with the wall must be avoided. Further, a large diameter-to-height ratio that still allows a high reduction degree should be used from the viewpoint of decreased heat loss. This may require the incorporation of multiple burners and solid feeding ports on the reactor roof.

REFERENCES

[1] Sohn, H. Y. *Steel Times International*, 2007, May/June, 68.
[2] Tacke K. H. and R. Steffen: *Stahl und Eisen*, 2004, 124 (4), 45.
[3] IISI, *Steel Statistical Yearbook* 2007, http://www.worldsteel.org/pictures/publicationfiles/SSY2007.pdf.
[4] Lockwood Greene Technologies: *Ironmaking Process Alternatives Screening Study*, Report LG Job No. 010529.01, U.S. Department of Energy, 2000.
[5] Brent, D., P. L. J. Mayfield, and T. A. Honeyands: In *International Conference on Alternative Routes of Iron and Steelmaking (ICARISM '99)*, pp. 111, eds. V. N. Misra and R. J. Holmes, The Australasian Institute of Mining and Metallurgy, Victoria, Australia, 1999.
[6] Husain, R., S. Sneyd, and P. Weber: In *International Conference on Alternative Routes of Iron and Steelmaking (ICARISM '99)*, p. 123, eds. V. N. Misra and R. J. Holmes, The Australasian Institute of Mining and Metallurgy, Victoria, Australia, 1999.
[7] Macauley, D. *Steel Times International*, 1997, 21 (1), 20.
[8] Sohn, H. Y., M. E. Choi, Y. Zhang, and J. E. Ramos: *Iron & Steel Technology (AIST Trans.)*, 2009, 6(8), 158.
[9] Sohn, H. Y., M. E. Choi, Y. Zhang, and J. E. Ramos: In *Energy Technology Perspectives: Carbon Dioxide Reduction and Production from Alternative Sources*, ed. by Neelameggham, N. R. Reddy, R. G. Belt, C. K. and Vidal, E. E. (also available in Collected Proceedings CD, TMS 2009 Annual Meeting), p. 93, TMS, Warrendale, PA, 2009.
[10] Sohn H. Y. and M. E. Choi: In *Collected Proceedings: Supplemental Proceedings: Vol. 1 Materials Processing and Properties, International Symposium on High-Temperature Metallurgical Processing*, p. 347, 139th TMS Annual Meeting, Seattle, WA, ed. by

Drelich, J. Hwang, J.-Y. Jiang, T. and Downey, J. TMS, Warrendale, PA, 2010.

[11] USGS: *2005 Minerals Yearbook - Iron Ore*, (*Report U.S. Department of the Interior, U.S. Geological Survey*), 2007.

[12] Pinegar, H. K., M. S. Moats, and H. Y. Sohn: *Steel Res. Int'l.*, 2011, 82 (8), 951. http://onlinelibrary.wiley.com/doi/10.1002/srin.201000288/pdf.

[13] Pinegar, H. K., M. S. Moats, and H. Y. Sohn: *Ironmaking Steelmaking*, 2012, 39, 398. www.ingentaconnect.com/content/maney/ias.

[14] Pinegar, H. K., M. S. Moats, and H. Y. Sohn: *Ironmaking Steelmaking*, 2013, 40, 32-43. http://www.ingentaconnect.com/content/maney/ias.

[15] Pinegar, H. K., M. S. Moats, and H. Y. Sohn, *Ironmaking Steelmaking*, 2013, 40, 44-49. http://www.ingentaconnect.com/content/maney/ias.

[16] Fan, D., Y. Mohassab, M. Elzohiery, and H. Y. Sohn: *Metall. Mater. Trans. B*, 2016, 47B (3), 1669-1680. doi:10.1007/s11663-016-0603-3.

[17] D.-Q. Fan, M. Elzohiery, Y. Mohassab and H. Y. Sohn, "Rate-Enhancement Effect of CO in Magnetite Concentrate Particle Reduction by H_2+CO Mixtures," *Ironmaking Steelmaking*, 2013, accepted for publication. https://doi.org/10.1080/03019233.2021.1915645.

[18] Fan, D. Q., H. Y. Sohn, and Mohamed Elzohiery, *Metall. Mater. Trans. B*, 2017, 48B, 2677-2684.

[19] Elzohiery, M., H. Y. Sohn, and Y. Mohassab, *Steel Research International*, 2017, 88 (2), 1600133 (14 pp.). http://dx.doi.org/10.1002/srin.201600133.

[20] M. Elzohiery, D.-Q. Fan, Y. Mohassab, and H. Y. Sohn, "Kinetics of Hydrogen Reduction of Magnetite Concentrate Particles at 1623 – 1873 K Relevant to Flash Ironmaking," *Ironmaking Steelmaking*, accepted for publication. https://doi.org/10.1080/03019233.2020.1819942.

[21] D.-Q. Fan, M. Elzohiery, Y. Mohassab, and H. Y. Sohn, "The Kinetics of Carbon Monoxide Reduction of Magnetite Concentrate Particles through CFD Modeling," *Ironmaking Steelmaking*, accepted for publication; https://doi.org/10.1080/03019233.2020.1861857.

[22] Wang H. and H. Y. Sohn, *Steel Research International*, 2012, 83, 903–909: http://dx.doi.org/10.1002/srin.201200054.

[23] Chen, F., Y. Mohassab, T. Jiang, and H. Y. Sohn, "Hydrogen Reduction Kinetics of Hematite Concentrate Particles Relevant to a Novel Flash Ironmaking Process," *Metall. Mater. Trans. B*, 2015, 46B (3), 1133-1145. http://link.springer.com/article/10.1007/s11663-015-0332-z.

[24] Chen, F., Y. Mohassab, S. Zhang, and H. Y. Sohn, "Kinetics of the Reduction of Hematite Concentrate Particles by Carbon Monoxide Relevant to a Novel Flash Ironmaking Process," *Metall. Mater. Trans. B*, 2015, 46B (4), 1716-1728.

[25] Fan, D. Q., H. Y. Sohn, Y. Mohassab, and M. Elzohiery, *Metall. Mater. Trans. B*, 2016, Vol. 47B (6), 3489-3500. doi:10.1007/s11663-016-0797-4; http://link.springer.com/article/10.1007/s11663-016-0797-4.

[26] Elzohiery, M., D. Q. Fan, Y. Mohassab, and H. Y. Sohn, "Experimental Investigation and Computational Fluid Dynamics Simulation of the Magnetite Concentrate Reduction Using Methane-Oxygen Flame in a Laboratory Flash Reactor," *Metall. Mater. Trans. B*, 2020, 51B, 1003–1015. https://doi.org/10.1007/s11663-020-01809-9.

[27] Elzohiery, M. *Flash Reduction of Magnetite Concentrate Related to a Novel Flash Ironmaking Process*, PhD Dissertation, University of Utah, 2018.

[28] Fan, D. Q. *Computational Fluid Dynamics analysis and Design of Flash Ironmaking Reactors*, PhD Dissertation, University of Utah, 2019.

[29] Abdelghany, D.-Q. Fan, M. Elzohiery, and H. Y. Sohn, *Steel Res. Int.*, 2019, 90 (9), 1900126 (10 pp.). https://doi.org/10.1002/srin.201900126.

[30] Jones W. P. and R. P. Lindstedt, *Combust. Flame*, 1988, 73(3), 233-249.

Reviewed by In-House

BIOGRAPHICAL SKETCH

Hong Yong Sohn

Affiliation: Department of Materials Science & Engineering, University of Utah

Education: PhD Chem. Eng., University of California – Berkeley

Business Address: 135 S 1460 E, Salt Lake City, Utah 84112

Research and Professional Experience:
Professor Sohn has developed a rate law called "Sohn's Law of Fluid-Solid Reactions." He also has extensive experience on managing multi-million-dollar research programs, including the Utah State Center of Excellence for Adv. Pyrometallurgical Technology (1988-93, $1.66MM) and the AISI/DOE project on Flash Ironmaking Technology (FIT) totaling $15 million over 3 successive phases.

Professional Appointments:
9/74 – Asst., Assoc., Full, Distinguished Professor, Materials Science & Engineering, University of Utah, Salt Lake City, UT
4/73 – 9/74, Research Engineer, Du Pont, Engineering Technology Lab, Wilmington, DE
2/71 – 4/73, Post-Doctoral Associate & Part-Time Lecturer, State Univ. of New York, Buffalo, NY
9/66 – 11/70, Research Assistant, Univ. of California, Berkeley, CA
Honorary Professor of Metallurgy, Department of Metallurgy, Kunming University of Science and Technology, Kunming, Yunnan, China
Honorary Professor, Anhui University of Technology, Maanshan, Anhui, China
Advisor, LS-Nikko Copper Inc., Ulsan, Korea
Advisor, Korea Institute of Geoscience and Mineral Resources (KIGAM)

Honors:

Distinguished Professor, 2014, University of Utah; 2014 Educator Award from TMS; Distinguished Scholarly and Creative Research Award, 2012 from the Univ. of Utah; Billiton Gold Medal, 2012 from The Inst. of Materials, Minerals and Mining in the U.K.; TMS 2009 Fellow Award TMS; In 2006 TMS honored Dr. Sohn with the "Sohn International Symposium on Advanced Processing of Metals and Materials." TMS Extraction and Processing Sci. Award (1990, 1994, 1999 and 2007, first four-time winner); 2001 James Douglas Gold Medal Award (AIME); Fellow Award, KAST (Korean Academy of Science and Technology), 1998; TMS Mathewson Gold Medal Award; TMS Extractive Metallurgy Lecturer Award (1990); Fulbright Distinguished Lecturer 1983; Camille and Henry Dreyfus Foundation Teacher-Scholar Award (1977).

Publications from the Last 3 Years: (out of 5 monographs, 17 edited books, 26 book chapters, some 570 papers, 4 patents)

(1) Sohn, H. Y. *Fluid–Solid Reactions*, Elsevier, Cambridge, MA 02139, 536 pp., 2020.

(2) Sohn, H. Y. Mohamed Elzohiery and De-Qiu Fan, "Development of Flash Ironmaking Technology," Chapter 2 in *Advances in Engineering Research*, vol. 26, V. M. Petrova, ed., ISBN: 987-1-715-5 (eBook); ISSN:2163-3932, Nova Science Publishers, Hauppauge, New York, 2019, pp. 23-106.

(3) Sohn H. Y. and A. Murali, "Plasma Synthesis of Advanced Metal Oxide Nanoparticles and Their Applications as Transparent Conducting Oxide Thin Films," *Molecules,* 26 (5), 1456 (2021). https://doi.org/10.3390/molecules26051456.

(4) Fan, D. Q., M. Elzohiery, Y. Mohassab and H. Y. Sohn, "Rate-Enhancement Effect of CO in Magnetite Concentrate Particle Reduction by H_2 + CO Mixtures," *Ironmaking Steelmaking.* https://doi.org/10.1080/03019233.2021.1915645.

(5) Sohn, H. Y., D. Q. Fan, and A. Abdelghany, "Design of Novel Flash Ironmaking Reactors for Greatly Reduced Energy Consumption and

CO$_2$ Emissions," *Metals,* https://www.mdpi.com/2075-4701/11/2/332/pdf.

(6) Ghadi, Z., M. S. Valipour, S. M. Vahedi, and H. Y. Sohn, "A Review on the Modeling of Gaseous Reduction of Iron Oxide Pellets," *Steel Res. Int.*, 91 (1), 1900270 (16 pp.) (2020). https://doi.org/10.1002/srin.201900270.

(7) Liu, M., D. Zhao, W. Zhai, J. Yang, B. Yang, H. Y. Sohn, and B. Xu, "Rapid preparation and properties investigation of TinO2n-1@C core-shell nanoparticles," *J. Alloys and Compounds*, 816, Article 152516 (2020). https://doi.org/10.1016/j.jallcom.2019.152516.

(8) Sarkar, R., B. P. Nash, and H. Y, Sohn, *"Interaction of magnesia-carbon refractory with ferrous oxide under the conditions of the novel flash ironmaking technology (FIT),"* Ceram. Int., https://reader.elsevier.com/reader/sd/pii/S027288421933408X?token=BFD1CCB425661078E2F0923B944B2D0739E7D8E9B6A14B3928C3BB995F9AA8A77CB1114B63E08F75862337C3FC3C56F2 Available online 28 November 2019.

(9) Sohn, H. Y. "Energy Consumption and CO2 Emissions in Ironmaking and Development of a Novel Flash Technology," *Metals,* 10 (1), 54 (2020). https://doi.org/10.3390/met10010054.

(10) Sohn, H. Y. "Principles and Applications of Mathematical and Physical Modelling of Metallurgical Processes," *Mineral Processing and Extractive Metallurgy* (IMM Transactions Section C), 129, 117-144 (2020). https://doi.org/10.1080/25726641.2019.1706376.

(11) Sohn H. Y. and S. Roy, "Fluid-Solid Reaction Kinetics for Solids of Non-Basic Geometries: Application of the Law of Additive Times in Combination with the Shape-Factor Method," *Metall. Mater. Trans. B*, 51, 601-610 (2020). https://doi.org/10.1007/s11663-020-01772-5.

(12) Elzohiery, M., D. Q. Fan, Y. Mohassab, and H. Y. Sohn, "Experimental Investigation and Computational Fluid Dynamics Simulation of the Magnetite Concentrate Reduction Using Methane-Oxygen Flame in a Laboratory Flash Reactor," *Metall. Mater. Trans. B,* 51, 1003–1015 (2020). https://doi.org/10.1007/s11663-020-01809-9.

(13) Sohn H. Y. and S. Roy, "Fluid-Solid Reaction Kinetics for Solids of Non-Basic Geometries: Comparison of the Sohn-Wall Method and the Shape Factor Method," *Ind. Eng. Chem. Res.*, 59, 5720–5724 (2020).

(14) O'Malley, K., F. Donat, K. J. Whitty, and H. Y. Sohn, "Scalable Preparation of Bimetallic Cu/Ni-based Oxygen Carriers for Chemical Looping," *Energy and Fuels*, 34 (9), 11227-11236 (2020). https://doi.org/10.1021/acs.energyfuels.0c01884.

(15) Zhao, H., Z. Song, H. Gao, B. Li, T. Hu, F. Liu, and H. Y. Sohn, "The structure-directing role of graphene in composites with porous FeOOH nanorods for Li ion batteries," *RSC Advances*, 10(68), 41403-41409 (2020).

(16) Shamsuddin M. and H. Y. Sohn, "Physical Chemistry of the Aqueous Processing of Inorganic Materials – A Review," *Trends in Physical Chemistry,* 20, 97–111 (2020).

(17) Elzohiery, M., D. Q. Fan, Y. Mohassab, and H. Y. Sohn, "Kinetics of Hydrogen Reduction of Magnetite Concentrate Particles at 1623 – 1873 K Relevant to Flash Ironmaking," *Ironmaking Steelmaking*. https://doi.org/10.1080/03019233.2020.1819942.

(18) Abdelghany, A., D. Q. Fan, and H. Y. Sohn, "Novel Flash Ironmaking Technology Based on Iron Ore Concentrate and Partial Combustion of Natural Gas: A CFD Study," *Metall. Mater. Trans. B*, 51, 2046–2056 (2020).

(19) Zhao, H., J. Wang, F. Liu, and H. Y. Sohn, "Experimental study on flow zone distribution and mixing time in a Peirce-Smith copper converter," *Int. J. Minerals, Metall. Materials*, accepted.

(20) Zhao, H., J. Wang, F. Liu, and H. Y. Sohn, "Experimental study on bubble distribution and splashing in a Peirce-Smith copper converter," *Metall. Mater. Trans, B,* 52 (1), 440-450 (2021). https://doi.org/10.1007/s11663-020-02030-4.

(21) Zhao, H., F. Liu, M. Xie, W. Liu, and H. Y. Sohn, "Recycling and Utilization of Spent Potlining by Different High Temperature Treatments," *J. Cleaner Production*, 289, 125704 (2021). https://doi.org/10.1016/j.jclepro.2020.125704.

(22) Fan, D. Q., M. Elzohiery, Y. Mohassab, and H. Y. Sohn, "The Kinetics of Carbon Monoxide Reduction of Magnetite Concentrate Particles through CFD Modeling," *Ironmaking Steelmaking*, https://doi.org/10.1080/03019233.2020.1861857.

(23) Murali, A., P, Sarswat, and H. Y. Sohn, "Enhanced Photocatalytic Activity and Photocurrent Properties of Plasma-Synthesized Indium-Doped Zinc Oxide Nanopowder," *Materials Today Chemistry*, 11, 60-68 (2019).

(24) Murali A. and H. Y. Sohn, "Photocatalytic Property of Plasma-Synthesized Alumina-Doped Zinc Oxide Nanopowder," *J. Nanoscience and Nanotechnology*, 19 (8), 4377-4386 (2019).

(25) Murali, A., H. Y. Sohn and P. K. Sarswat, "Plasma-Assisted Chemical Vapor Synthesis of Aluminum-Doped Zinc Oxide Nanopowder and Synthesis of AZO Films for Optoelectronic Applications," *J. Electronic Materials*, 48 (4), 2531-2542 (2019). https://doi.org/10.1007/s11664-019-06926-z.

(26) Sohn, H. Y. "Review of Fluid-Solid Reaction Analysis: Part 1. Single Nonporous Reactant Solid," *Can. J. Chem. Eng.*, 97, 2061-2067 (2019). https://doi.org/10.1002/cjce.23469.

(27) Sohn, H. Y. "Review of Fluid-Solid Reaction Analysis: Part 2. Single Porous Reactant Solid," *Can. J. Chem. Eng.*, 97, 2068-2076 (2019). https://doi.org/10.1002/cjce.23468.

(28) Sohn, H. Y. "Review of Fluid-Solid Reaction Analysis: Part 3. Complex Fluid-Solid Reactions," *Can. J. Chem. Eng.*, 97, 2326-2332 (2019). https://doi.org/10.1002/cjce.23475.

(29) Murali, A., Y. P. Lan, and H. Y. Sohn, "Effect of Oxygen Vacancies in Non-Stoichiometric Ceria on Its Photocatalytic Properties," *Nano-Structures & Nano-Objects*, 18, 100257 (2019).

(30) Sohn, H. Y., S. Roy, and D. Q. Fan, "Fluid-Solid Reaction Kinetics for Solids of Nonbasic Geometries and Determination of the Appropriate Shape Factors," *Metall. Mater. Trans. B*, 50, 2037-2046 (2019). https://doi.org/10.1007/s11663-019-01592-2.

(31) Sarkar R. and H. Y. Sohn, "Interaction of Iron with Alumina Refractory under Flash Ironmaking Conditions," *Metall. Mater.*

Trans. B, 50, 2063–2076 (2019). https://link.springer.com/content/pdf/10.1007%2Fs11663-019-01610-3.pdf.

(32) Sarkar R. and H. Y. Sohn, "Interaction of pure alumina refractory with FeO-SiO2 and FeO-SiO2-CaO slags relevant to the novel Flash Ironmaking Technology (FIT)" *Steel Research International, Steel Research International,* 90 (9), 1900104 (13 pp.) (2019). https://doi.org/10.1002/srin.201900104.

(33) Shamsuddin M. and H. Y. Sohn, "Constitutive Topics in Physical Chemistry of High-Temperature Nonferrous Metallurgy – A Review: Part 1. Sulfide Roasting and Smelting," *JOM,* 71 (9), 3253–3265 (2019).

(34) Shamsuddin M. and H. Y. Sohn, "Constitutive Topics in Physical Chemistry of High-Temperature Nonferrous Metallurgy – A Review: Part 2. Reduction and Refining," *JOM,* 71 (9), 3266–3276 (2019).

(35) Sarkar R. and H. Y. Sohn, "Interaction of ferrous oxide with alumina refractory under flash ironmaking conditions," *Ceramics International,* 45 (12), 15417-15428 (2019).

(36) Abdelghany, A., D. Q, Fan, M, Elzohiery, and H. Y. Sohn, "Experimental Investigation and Computational Fluid Dynamics Simulation of a Novel Flash Ironmaking Process Based on Partial Combustion of Natural Gas in a Large-Scale Bench Reactor," *Steel Res. Int.,* 90 (9), 1900126 (10 pp.) (2019). https://doi.org/10.1002/srin.201900126.

(37) Sarkar, R., B. P. Nash, and H. Y, Sohn, "Interaction of magnesia-carbon refractory with metallic iron under flash ironmaking conditions," *J. European Ceram. Soc.*, 40(2), 529-541 (2019).

(38) Shamsuddin M. and H. Y. Sohn, "Constitutive Topics in Physical Chemistry of Ironmaking and Steelmaking – A Review," *Trends in Physical Chemistry,* 19, 33-50 (2019).

(39) Lan Y. P. and H. Y. Sohn, "Nanoceria Synthesis in Molten KOH-NaOH Mixture: Characterization and Oxygen Vacancy Formation," *Ceramics International,* 44 (4), 3847-3855 (2018). https://doi.org/10.1016/j.ceramint.2017.11.172.

(40) Xu, B., D. Zhao, H. Y. Sohn, Y. Mohassab, B. Yang, Y. Lan, and J. Yang, "Flash Synthesis of Magnéli Phase ($TinO_{2n-1}$) Nanoparticles by Thermal Plasma Treatment of H_2TiO_3," *Ceramics International*, 44 (4), 3929-3936 (2018); 45 (5) 6602 (2019). https://doi.org/10.1016/j.ceramint.2017.11.184.

(41) Lan Y. P. and H. Y. Sohn, "Effect of Oxygen Vacancies and Phases on Catalytic Properties of Hydrogen-Treated Nanoceria Particles," *Materials Research Express*, 5, 035501 (2018). https://doi.org/10.1088/2053-1591/aaaff4.

(42) Sarkar R. and H. Y. Sohn, "Interactions of Alumina and Magnesia Based Refractories with Iron Melts and Slags - A Review," *Metall. Mater. Trans. B*, 49, 1860–1882 (2018). https://link.springer.com/article/10.1007/s11663-018-1300-1.

(43) Murali and H. Y. Sohn, "Plasma-Assisted Chemical Vapor Synthesis of Indium Tin Oxide (ITO) Nanopowder and Hydrogen-Sensing Property of ITO Thin Film," *Materials Research Express*, 5 (6), 065045 (2018).

(44) Murali A. and H. Y. Sohn, "Photocatalytic Properties of Plasma-Synthesized Zinc Oxide and Tin-Doped Zinc Oxide (TZO) Nanopowders and Their Applications as Transparent Conducting Films," *J. Materials Science: Materials in Electronics*, 29 (17), 14945-14959 (2018).

(45) Lan, Y. P., H. Y. Sohn, A. Murali, J. Li, and C. Chen, "The Formation and Growth of CeOCl Crystals in a Molten KCl-LiCl Flux," *Applied Physics A*, 124 (10), 702 (2018).

(46) Elzohiery, M., H. Y. Sohn, and Y. Mohassab, "Kinetics of Hydrogen Reduction of Magnetite Concentrate Particles in Solid State Relevant to Flash Ironmaking," *Steel Research International*, 88 (2), 1600133 (14 pp.) (2017). http://dx.doi.org/10.1002/srin.201600133.

(47) Lan, Y. P., H. Y. Sohn, Y. Mohassab, Q. Liu, and B. Xu, "Nanoceria Synthesis in the KCl-LiCl Salt System: Crystal Formation and Properties," *J. Am. Ceram. Soc.*, 100, 1863–1875 (2017). doi: 10.1111/jace.14747.

(48) Bronson, T. M., N. Y. Ma, L. Z. Zhu, and H. Y. Sohn, "Oxidation and Condensation of Zinc Fume from $Zn-CO_2-CO-H_2O$ Streams Relevant to Steelmaking Off-Gas Systems," *Metall. Mater. Trans. B,* 48B (2), 908-921 (2017).

(49) Sohn H. Y. and D. Q. Fan, "On the Initial Rate of Fluid-Solid Reactions," *Metall. Mater. Trans. B,* 48B, 1827-1832 (2017). DOI: 10.1007/s11663-017-0940-x.

(50) Morales-Estrella, R., J. Ruiz, N. Ortiz-Lara, Y. Mohassab, and H. Y. Sohn, "Effect of Mechanical Activation on the Hydrogen Reduction Kinetics of Magnetite Concentrate," *J. Thermal Analysis and Calorimetry*, 130 (2), 713–720 (2017).

(51) Fan, D. Q., H. Y. Sohn, and M. Elzohiery, "Analysis of the Reduction Rate of Hematite Concentrate Particles by H2 or CO in a Drop-Tube Reactor through CFD Modeling," *Metall. Mater. Trans. B,* 48B, 2677-2684 (2017). http://rdcu.be/ybnM; https://link.springer.com/content/pdf/10.1007%2Fs11663-017-1053-2.pdf.

(52) Sohn, H. Y. "Book Review: Physical Chemistry of Metallurgical Processes," *Advanced Science News*, http://www.advancedsciencenews.com/book-review-physical-chemistry-metallurgical-processes/, posted September 20, 2017.

(53) Lan, Y. P., H. Y. Sohn, Y. Mohassab, Q. Liu, and B. Xu, "Properties of Stable Nonstoichiometric Nanoceria Produced by Thermal Plasma," *J. Nanoparticle Research*, 19, 281 (2017). https://doi.org/10.1007/s11051-017-3984-6.

(54) Lan, Y. P., H. Y. Sohn, and Q. Liu, "Plasma-Assisted Synthesis of Non-Stoichiometric Nanoceria Powder from Cerium Carbonate Hydroxide ($CeCO_3OH$)," *Metall. Mater. Eng.*, 23 (3), 213-225 (2017).

In: Computational Fluid Dynamics
Editor: James S. Hutchinson

ISBN: 978-1-53619-756-3
© 2021 Nova Science Publishers, Inc.

Chapter 3

MODELING AND COMPUTATIONAL FLUID DYNAMICS (CFD) SIMULATION OF CO_2 ABSORPTION USING MONO-ETHANOL AMINE (MEA) SOLUTION IN A HOLLOW FIBER MEMBRANE (HFM) CONTACTOR

Ehsan Kianfar[*] *and Sajjad Golchin Khazari*[†]
Young Researchers and Elite Club, Gachsaran Branch,
Islamic Azad University, Gachsaran, Iran
Department of Mechanic Engineering, Qazvin Branch,
Islamic Azad University, Qazvin, Iran

ABSTRACT

In this study, computational fluid dynamics technique (CFD) was used to simulate a hollow fiber membrane (HFM) contactor for the absorption of carbon dioxide from the air by mono-ethanol amine (MEA). The

---

[*] Corresponding Author's Email: e-kianfar94@iau-arak.ac.ir.
[†#] Corresponding Author's Email: ehsan_kianfar2010@yahoo.com.

governing equations of this process include the equations of continuity, momentum and mass transfer in areas related to the shell membrane and tube and by taking into account the relevant boundary conditions and assumptions and using the software COMSOL, the equations were solved. By examining and comparing the results with experimental data presented in the literature, the accuracy of simulation results was confirmed. The influences of important parameters in the process including wetness of membrane, the volumetric flow rate of gas, liquid volumetric flow rate and temperature effects were studied. Results show that with increasing the volumetric flow rate of the liquid absorber, the mass transfer rate of carbon dioxide to absorption increases because of the carbon dioxide concentration gradient in the gas phase and liquid increases. Because of the reduced residence time in the membrane contactor, with increasing the volumetric flow rate of gas the amount of removed CO_2 in contact reduces so that with increasing the gas flow rate from 0.6 to 1.4, removed CO_2 reduces only 18 percent. The other factors are the solvent temperature and wetness. With increasing the solvent temperature, carbon dioxide removal efficiency is significantly increased. Wettability has a strong negative effect on the efficiency of the membrane contactor. Thus with increasing wettability, the efficiency of the membrane suddenly finds a sharp decrease. In some cases, increased wettability efficiency is close to zero.

Keywords: CO_2 capture, computational fluid dynamics (CFD), hollow fiber membrane, mono-ethanol amine (MEA), modeling, simulation

NOMENCLATURE

A	liquid–gas contact area [m^2]
a$_0$	initial concentration ratio of MEA; a$_0$ = nMEA/ (nMEA + nH$_2$O)
C	concentration [mol m^{-3}]
D	diffusion coefficient [m^2 s^{-1}]
DCO2	diffusion coefficient of CO2 [m^2 s^{-1}]
d$_h$	hydraulic diameter [m]
Dig e	effective membrane diffusion coefficient [m^2 s^{-1}]
Dig k	Knudsen diffusion coefficient of CO2 [m^2 s^{-1}]
Dig m	molecular diffusion coefficient of CO2 [m^2 s^{-1}]
DN2O	diffusion coefficient of N2O [m^2 s^{-1}]

do	pore mean diameter [m]
dp	pore equivalent diameter for membrane [m] gas [cm3 mol−1]
H	Henry's constant [kPa m^3 kmol^{-1}]
k−1	reverse first-order reaction rate constant [s^{-1}]
k2	second-order forward reaction rate constant [m^3 mol^{-1} s^{-1}]
kb	second-order forward reaction rate constant for base b [m^3 mol^{-1} s^{-1}]
KCO2	combined Henry's law and chemical equilibrium constant;
KCO2	KCO2 ′ H [kPa]
KCO2 ′	chemical equilibrium constant for a single chemical reaction [kmol m−3]
kex	external mass-transfer coefficient [m s^{-1}]
kg	mass-transfer coefficient in gas phase [m s^{-1}]
kL	mass-transfer coefficient in liquid phase [m s^{-1}]
km	mass-transfer coefficient in membrane phase [m s^{-1}]
L	fiber length of membrane module [m]
m	distribution coefficient between liquid and gas
M	mole weight [g mol^{-1}]
NCO2	CO2 absorption flux [mol m^{-2} s^{-1}]
P	pressure [kPa]
Q	flow rate [m^3 s^{-1}]
R	fiber inner radius [m]
$\bar{R}$	ideal gas constant [J mol^{-1} K^{-1}]
r	radial coordinate [m]
rCO2	CO 2 reaction rate [mol s^{-1}]
Re	Reynolds number
rMEA	MEA reaction rate [mol s^{-1}]
Sc	Schmidt number
Sh	Sherwood number
$\bar{T}$	dimensionless temperature
T	temperature [K]
$\bar{V}$	liquid mean velocity [m s−1]
VZ	liquid velocity in axial direction [m s−1]

X	excess Henry's coefficient
XCO2	mole fraction of chemically bound CO2 in the solution
yO2	O2 loss ratio [%] ν = molecular volume of
ΔP	breakthrough pressure [MPa]

Greek Symbols

[%] μ	viscosity of gas [Pa s]
[m] τ	tortuosity ς = membrane porosity
α	CO2 loading [mol CO2/mol MEA]
γ	surface tension [mN/m]
δ	membrane thickness
ε	Stockmayer potential parameter [J]
η	viscosity of liquid [mPa s]
ηCO2	CO2 removal efficiency
θ	contact angle [°]
κ	Boltzmann constant [J K−1]
ρ	density [kg m−3]
ϕ	membrane pore wetting ratio
φ	packing density
Φi	volume fraction of the ith solvent
ΩD	diffusion collision integral
Ωμ	viscosity collision integral

Subscripts

e	equilibrium
G	gas phase of membrane contactor
i	gas/liquid interface
L	liquid phase of membrane contactor
out	outlet of membrane contactor

Superscripts

    in   inlet of membrane contactor
    w   water

1. INTRODUCTION

Nowadays, increased greenhouse gases emissions into the atmosphere are one of the most important environmental issues. Among greenhouse gases, carbon dioxide is known as the most important greenhouse gas due to having the highest retention time and the greatest rate in the atmosphere. Due to global concerns on earth warming, the advanced practices of greenhouse gases separation (in particular carbon dioxide) are considerably paid into attention [1-3]. Carbon dioxide not only makes damages on the environment but also results in the thermal evaluation of natural gas in power plants. Thanks to acidic nature of CO_2, its presence in natural gas pipelines may cause to wear the equipment and pipeline. Such issues make removal of CO_2 from natural gas as well as the prevention of carbon dioxide emissions in the atmosphere much more important [4-6]. In some cases, the removal of carbon dioxide from natural gases is used to prevent catalyst poisoning in the industry [7-10]. As an example in ammonia production, carbon dioxide is removed from hydrogen since such carbon is a poison for ammonia production catalyst [11-14]. In high concentrations, carbon dioxide causes to wear the equipment and results in increased gas volume and thereby increases the costs of condensation and gas transfer. Also, carbon dioxide thermally devaluates the gas [15-18]. Also, carbon dioxide is an important greenhouse gas which is produced by human and industrial activities. Emissions of this gas into the air lead to increase the earth temperature [19-21]. According to the above-mentioned causes and to improve and reduce the costs of gas refinement equipment, it makes sense to remove carbon dioxide from industrial flows and household applications in particular natural gas (reduction up to 2 mole %) due to tremendous reserves of this vital material [22-25]. Membrane processes are of the latest separation methods where removal is done with lower energy consumption. Nowadays, these processes are known as a proper alternative for traditional processes of gas absorption with amine solution [26-28]. Membrane processes provide considerable advantages compared to absorption processes including lower energy consumption, lower operational costs, and less initial required investment [22]. No risk of fire and explosion is seen in such systems. They can treat the gas over the well while high pressure is so proper for membrane processes, being environmental-friendly [23]. Due to being light and small, these systems are suitable for the offshore works. They provide excellent flexibility to comply with flow rate changes and feed composition and are absent in chemical processes. They are also able to separate concurrently hydrogen sulfide, carbon dioxide, and water vapor. Also, no energy is wasted through heat exchange with the environment in such processes since they are often applied at ambient temperature [25-29]. The main purpose of this study was to develop and solve a 2D mathematical model for the absorption of CO_2 in hollow fiber membrane (HFM) contactor. Axial and radial diffusions inside the shell through the membrane and within the tube side of the membrane are considered in the model equations. Convection in the shell and tube, as well as a chemical reaction, are also considered. The equations of continuity, momentum and mass transfer are solved using the software COMSOL. The simulation aimed to predict the concentration of gas components in the membrane contactor. The influence of various process parameters on the mass transfer of CO_2 was investigated. Physical and chemical absorptions for "non-wetted mode," where the membrane pores are filled with the gas mixture, have been considered in this work. Chemical absorption is considered for the absorption of CO_2 in aqueous solutions of mono-ethanol amine (MEA). The model was then validated using experimental data obtained from the literature for the absorption of CO_2 in the amine aqueous solution.

2. MASS TRANSFER MODEL

The analyzed geometry is shown schematically in Figure 1. As can be seen, a fluid (carbon dioxide) moves through inside the fiber and another fluid (mono-ethanol-amine) moves outside the fiber in the opposite direction and inside the tube. Given the experimental work, the number of fibers is 500. The additional properties of the used geometry are presented in Table 1. As can be seen in this table, the useful length which is used in the simulation is 780 mm. The corresponding Mass transfer in Equation 1 is as follows [30]:

$$\nabla \cdot (-D_i C_i + C_i u) + \frac{\partial C_i}{\partial t} = R_i \quad (1)$$

Table 1. Specifications of the membrane module [30]

parameter	value
module I.D	0.02 m
fiber O.D	344 μm
fiber I.D	424 μm
fiber active length	0.78 m
number of fibers	500
packing density	22.47%
contactor area (inner)	0.39 m²

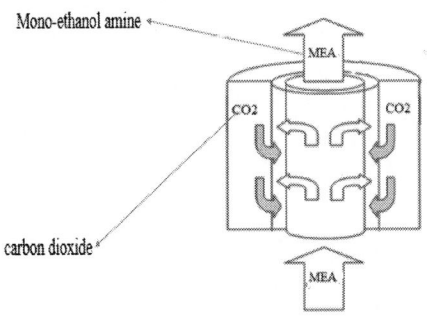

Figure 1. Schematic of carbon dioxide absorption in hollow fibers by monoethanolamide [25].

2.1. Numerical Procedure

Triangular mesh elements used for the evaluation of gas behavior in the HFM contactor are illustrated in Figure 2 in which r1 and r2 represent the radius of the shell side and free surface, respectively.

- Model equation

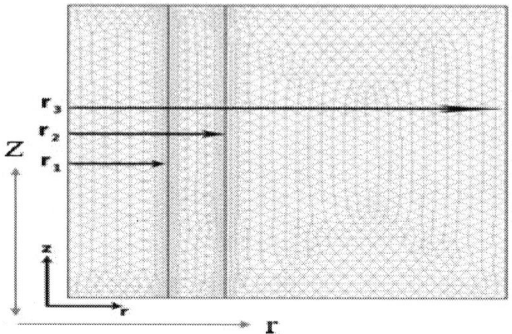

Figure 2. Model domain and FE model used for simulation.

By applying Navies–Stokes equation, velocity distribution in the tube side can be derived, but for developing velocity profile in the tube side, a fully developed Newtonian laminar flow is assumed the continuity equation for each component can be expressed as follows [30]:

$$V_{z-tube} = 2u\left[1-\left(\frac{r}{r_1}\right)^2\right] \quad [30] \quad (2)$$

$$\frac{\partial C_i}{\partial t} = -(\nabla . C_i V) - (\nabla . J_i) + R_i \quad [30] \quad (3)$$

- Tube side equations

In the tube side of HFMC, there is no chemical reaction, in addition to the fact that convective mass transfer in the radial direction is not considered in comparison with the axial direction. Consequently Eq. (3) can be written as follows [30]:

$$D_{CO2-tube}\left[\frac{\partial^2 C_{co2-tube}}{\partial r^2}+\frac{1}{r}\frac{\partial C_{co2-tube}}{\partial r}+\frac{\partial^2 C_{co2-tube}}{\partial z^2}\right]=V_{z-tube}\frac{\partial C_{co2}}{\partial z} \quad [30] \quad (4)$$

At $r=r_1, C_{co2-tube}=C_{co2-membrane}$ [30] (5)

At $r=0, \dfrac{\partial C_{co2-tube}}{\partial r}=0$ [30] (6)

At z = 0, convective flux [30] (7)

At z = L $C_{co2-tube}=C_{co2,0}$, [30] (8)

- Shell side equations

The steady-state continuity equation in the shell side is expressed as follows [30]:

$$D_{Co2-shell}\left[\frac{\partial^2 C_{co2-shell}}{\partial r^2}+\frac{1}{r}\frac{\partial C_{co2-shell}}{\partial r}+\frac{\partial^2 C_{co2-shell}}{\partial z^2}\right]=V_{z-shell}\frac{\partial C_{co2-shell}}{\partial z}-R_{co2} \quad [30] \quad (9)$$

$$-\nabla.\eta\left(\nabla V_{z-shell}+\left(\nabla V_{z-shell}\right)^T\right)+\left(\rho V_{z-shell}.\nabla\right)V_{z-shell}+\nabla\rho\nabla.V_{z-shell}=0 \quad [30] \quad (10)$$

$$r_3=\left(\frac{1}{1-\phi}\right)^{1/2}r_2 \quad [30] \quad (11)$$

$$1-\phi=\frac{nr_2^2}{R^2} \quad [30] \quad (12)$$

At $r=r_3, \dfrac{\partial C_{co2-shell}}{\partial r}$ [30] (13)

At $r = r_2, C_{co2-shell} = C_{co2-membrane} \times m$ [30] (14)

At $z = 0, C_{co2-shell} = 0, V_{z-shell} = V_0$ [30] (15)

At $z = l$, Convective flux, $p = p_{atm}$ [30] (16)

- Membrane equations

Transport of CO₂ into the membrane pores is accomplished just via a diffusion mechanism. Steady-state and no reaction assumptions will simplify the continuity equation as follows [30]:

$$D_{CO2-membrane}\left[\frac{\partial^2 C_{co2-membrane}}{\partial r^2} + \frac{1}{r}\frac{\partial C_{co2-membrane}}{\partial r} + \frac{\partial^2 C_{co2-membrane}}{\partial z^2}\right] = 0 \text{ [30]} \quad (17)$$

At $r = r_1, C_{co2-membrane} = C_{co2-membrane}$ [30] (18)

At $r = r_2, C_{co2-membrane} = \dfrac{C_{co2-membrane}}{m}$ [30] (19)

- Absorbent equations

The steady-state continuity equation for potassium threonine absorber is written as follows [30]:

$$D_{i-shell}\left[\frac{\partial^2 C_{i-shell}}{\partial r^2} + \frac{1}{r}\frac{\partial C_{i-shell}}{\partial r} + \frac{\partial^2 C_{i-shell}}{\partial z}\right] = V_{z-shell}\frac{\partial C_{i-shell}}{\partial z} - R_i \text{ [30]} \quad (20)$$

At $r = r_3, \dfrac{\partial C_{absorbent-shell}}{\partial r} = 0$ [30] (21)

At $r = r_2$, $\dfrac{\partial C_{absorbent-shell}}{\partial r} = 0$ [30] (22)

At $z = l$, $C_{absorbent-shell} = C_{absorbent,0}$ [30] (23)

At $z = l$, Convective flux [30]

2.2. Reaction Mechanism of CO₂ with MEA

The reaction of CO_2 with MEA can be described by a two-step zwitterion mechanism, which was first proposed by Caplow and re-introduced by Danckwerts. The first step of the reaction is to form a zwitterion as an intermediate, which can be expressed as [30]

$$CO_2 + R_1R_2NH \underset{k_{-1}}{\overset{k_{2,amine}}{\rightleftharpoons}} R_1R_2NH^+COO^- \text{ [30]} \quad (24)$$

$$R_1R_2NH^+COO^- + B \xrightarrow{K_b} R_1R_2NCOO^- + BH^+ \text{ [30]} \quad (25)$$

$$r_{CO2} = \dfrac{k_{2,MEA}[MEA]([CO_2]-[CO_2]_e)}{1 + \dfrac{k_{-1}}{k_{H2O}[H_2O] + k_{MEA}[MEA]}} \text{ [30]} \quad (26)$$

The corresponding reaction rate constants in Equation 26 are as follows [30]:

$$k_{2,MEA}\left(m^3 kmol^{-1}s^{-1}\right) = 7.973 \times 10^{12} \exp\left[-\dfrac{6243}{T(K)}\right] \text{ [30]} \quad (27)$$

$$\frac{k_{2,MEA} k_{H2O}}{k_{-1}} \left(m^6 kmol^{-2} s^{-1} \right) = 1.1 \times 10^6 \exp\left[-\frac{3472}{T(K)} \right] \quad [30] \quad (28)$$

$$\frac{k_{2,MEA} k_{MEA}}{k_{-1}} \left(m^6 kmol^{-2} s^{-1} \right) = 1.563 \times 10^{14} \exp\left[-\frac{7544}{T(K)} \right] \quad [30] \quad (29)$$

$$[CO2]_e = K'_{CO2} [MEACOO^-] \left\{ \frac{a_0 \alpha}{[a_0(1-2\alpha)]^2} \right\} = \frac{K_{CO2} X_{CO2}}{H} \left\{ \frac{a_0 \alpha}{[a_0(1-2\alpha)]^2} \right\} \quad [30] \quad (30)$$

$$\ln K_{CO2} = A + \frac{B}{T} + Ca_0 \alpha + D\sqrt{a_0 \alpha} \quad [30] \quad (31)$$

The regressed parameters A-D is available in Gabrielsen's work.

3. RESULTS AND DISCUSSION

3.1. CO₂ Removal Using MEA Adsorbent through HFM

Figure 3 shows the meshed blocks which were used to assess gas transfer behavior in the membrane contactor. The meshes were generated by COMSOL software where the nodes were heterogeneous and unequal such that close to the walls, the finer meshes were used with 994 and 9410 nodes in the transverse direction and contactor length direction, respectively. Distribution of carbon dioxide concentration is shown in three sections including shell, tube, and membrane in Figure 4 where carbon dioxide is in line with contactor membrane and flows up from Z = 0.

As can be seen, the concentration of carbon dioxide is decreased along the shell direction since it penetrates the tube while in the tube section, the highest concentration of carbon dioxide is visible at the membrane contactor and absorbent solution. In doing so, due to the reactions done in the tube section between carbon dioxide and mono-ethanol-amine, the concentration of carbon dioxide is decreased and reaches zero at the center of the tube.

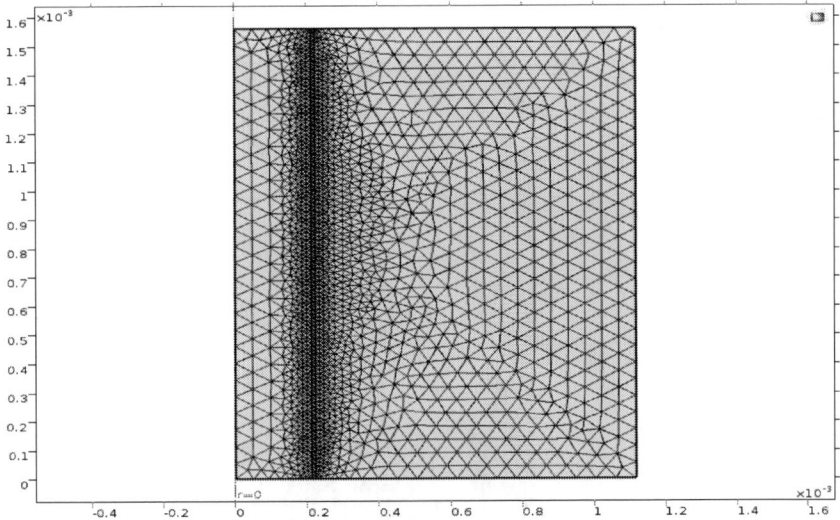

Figure 3. The classified elements of contactor membrane.

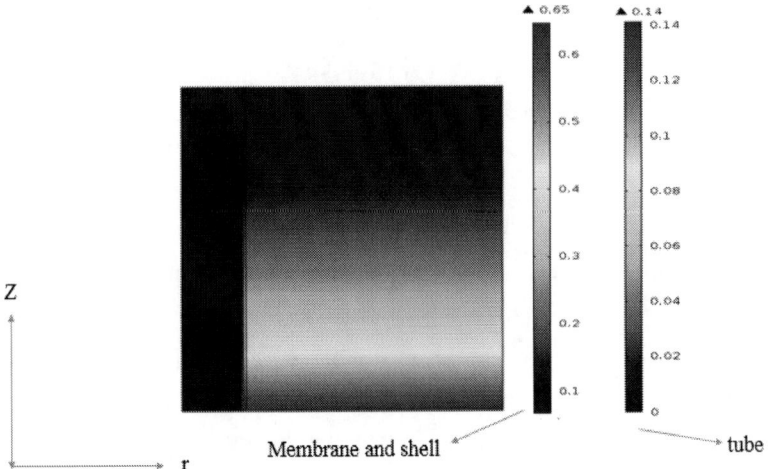

Figure 4. Distribution of concentration at three sections of contactor hollow fiber membrane.

3.2. Distribution of CO_2 Mass Transfer Flux in the Contactor Shell

Figure 5 shows the mass transfer flux at colored surfaces in the shell section and axial direction. It shows replacement and penetration in this area.

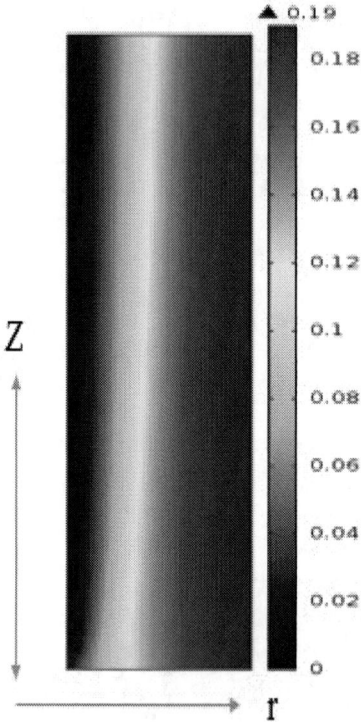

Figure 5. General flux distribution of carbon dioxide in the shell along the axial direction.

As can be seen, the lowest mass transfer is in the vicinity of contactor membrane and shell since the fluid velocity is decreased here and it has the null value at the contact surface. Therefore, the molecular influx is mass transfer mechanism with the lowest value since the mass transfer of replacement is decreased and reaches zero at the contactor membrane. As can be seen in this figure, there is no mass transfer in the axial direction at contactor membrane-shell.

3.3. Effects of Liquid Volumetric Flow Rate on CO_2 Absorption

The removal rate of carbon dioxide can be found through the following equation:

$$\% \text{ removal } CO_2 = (1 - \frac{C_{outlet}}{C_{inlet}}) \quad [30] \quad (32)$$

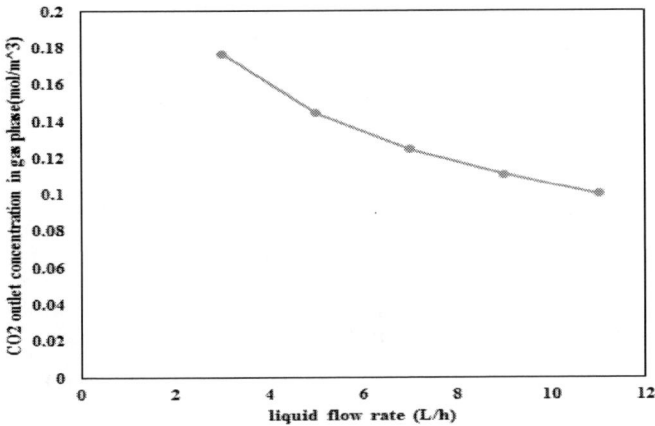

Figure 6. The relationship between output concentrations at the gas phase and the liquid volumetric flow rate.

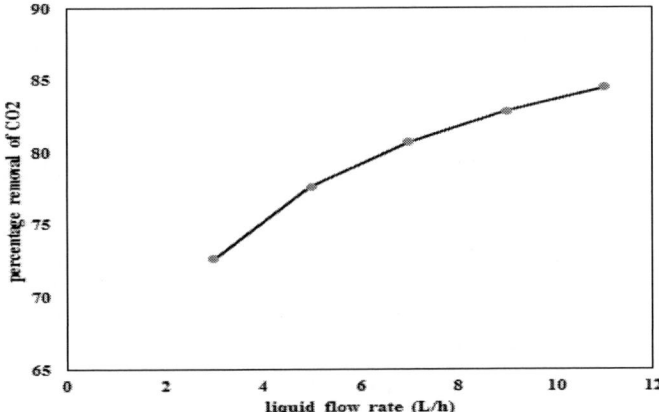

Figure 7. The relationship between carbon dioxide removal rate and liquid volumetric flow rate.

Figure 6 shows the output concentration of carbon dioxide in the gas phase at the shell section as a function of volumetric flow rate. Figure 7 depicts the carbon dioxide removal rate as a function of the liquid volumetric flow rate. As can be seen, the higher volumetric flow rate of the absorbent, the higher transfer rate of CO_2 mass to liquid absorbent due to increased carbon dioxide concentration gradient in gas and liquid phases. Also, with increased velocity in the liquid phase, the penetration factor in the liquid phase is increased, as well. In other words, mass transfer resistance is decreased and thereby the removal rate is enhanced and as a result, output concentration in the gas phase is lowered (Figure 6) consequently, removal range of carbon dioxide is increased Figure 7.

3.4. Effect of Gas Volumetric Flow Rate on Carbon Dioxide Flow Rate

Carbon dioxide removal rate in the gas phase of the contactor membrane is shown in Figure 8 for diverse volumetric flow rates. With increased gas volumetric flow rate, carbon dioxide removal rate in the contactor is decreased due to the lower time of stay in the contactor membrane.

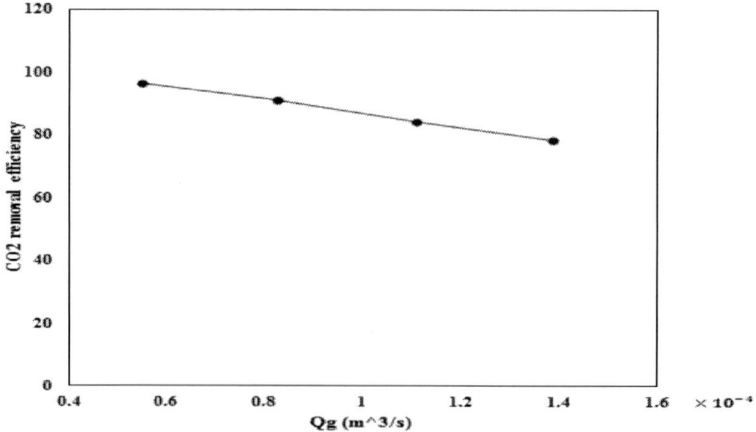

Figure 8. Carbon dioxide removal efficiency as a function of the gas volumetric flow rate.

Therefore, with increasing gas flow rate from 0.6 to 1.4, the CO_2 removal rate will be only 18% decreased. As a result, carbon dioxide absorption does not strongly follow the gas phase velocity. In this regard, as can be seen from Figure 7, carbon dioxide removal efficiency strongly follows liquid phase velocity while such a following is stronger in lower velocities and high liquid phase velocity, carbon dioxide absorption process is less dependent on liquid phase velocity. Therefore, it can be concluded from the results that the contactor membrane encompasses mass transfer controlling liquid phase in this process.

3.5. Effect of the Absorbent Temperature on the Removal Rate

Temperature is an effective and important operational parameter in the carbon dioxide absorption process among membranes. Figure 9 shows the effect of absorbent temperature on carbon dioxide removal efficiency. The effect of mono ethanol amine solvent temperature was investigated in the range of 300-350 Kelvin. As can be seen, with increased absorbent solvent temperature, carbon dioxide removal efficiency is considerably increased.

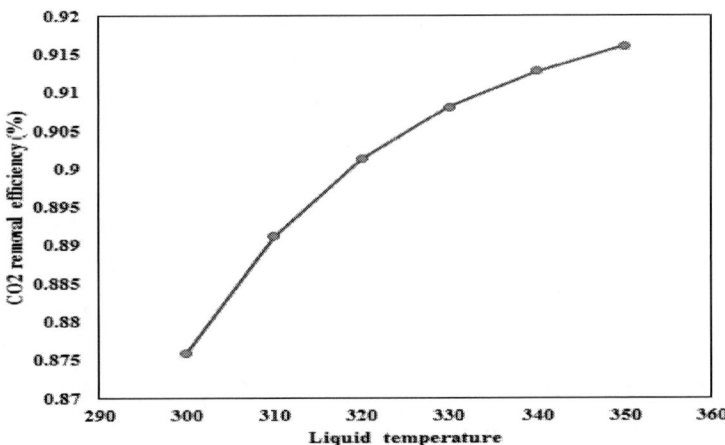

Figure 9. Effect of absorbent temperature on carbon dioxide removal efficiency.

This is because, with increased absorbent solvent temperature, penetration and velocity of reaction are increased in a solvent which causes to increase carbon dioxide removal range. Therefore, the absorbent solvent would better to enter the contactor membrane with a high temperature to reach a higher absorption process efficiency.

3.6. Effect of the Membrane Wettability on CO_2 Removal Efficiency at the Contactor Membrane

Figure 10 shows the effect of the membrane wettability on carbon dioxide removal efficiency at the contactor membrane. As can be seen in this figure, wettability has a powerful negative impact on carbon dioxide removal efficiency at the contactor membrane. Consequently, with increased wettability of the membrane, carbon dioxide removal efficiency will be suddenly and considerably decreased while the removal efficiency will approach the zero with increased wettability. For example, in the case of a membrane wettability of only 10%, the removal efficiency will be suddenly decreased from 86% to 20%. This may approve that absorption process efficiency strongly follows the membrane wettability since mass transfer factor is considerably decreased with higher wettability, increasing the mass transfer resistance substantially. Therefore, wettability is an important parameter in the carbon dioxide absorption process at the contactor membrane, which should be taken into account. In this case, the hydrophobic membrane should be used for which diverse methods exist.

Figure 11 shows the effect of wettability on carbon dioxide concentration along the tube. As can be seen, the higher the wettability, the lower carbon dioxide concentration. This is because CO_2 mass transfer factor to the absorption solvent is decreased and mass transfer resistance is increased. Also, by a wettability of 100%, carbon dioxide concentration reaches the zero, penetrating no carbon dioxide into the absorbent [31]

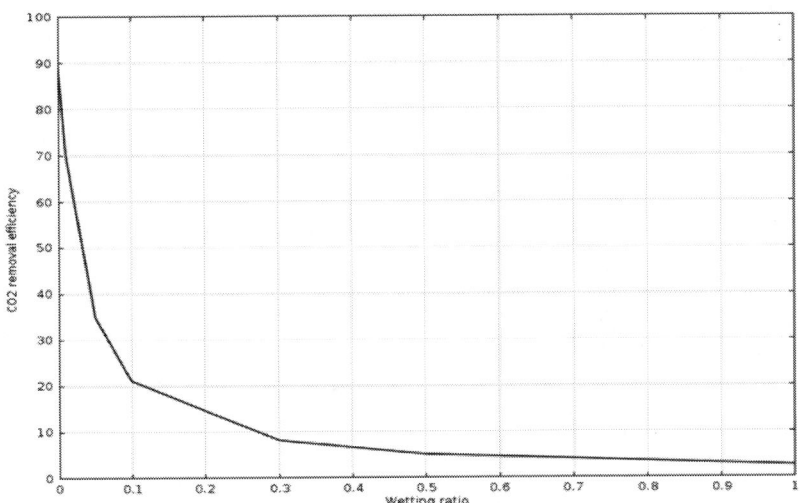

Figure 10. Carbon dioxide removal efficiency in terms of membrane wettability.

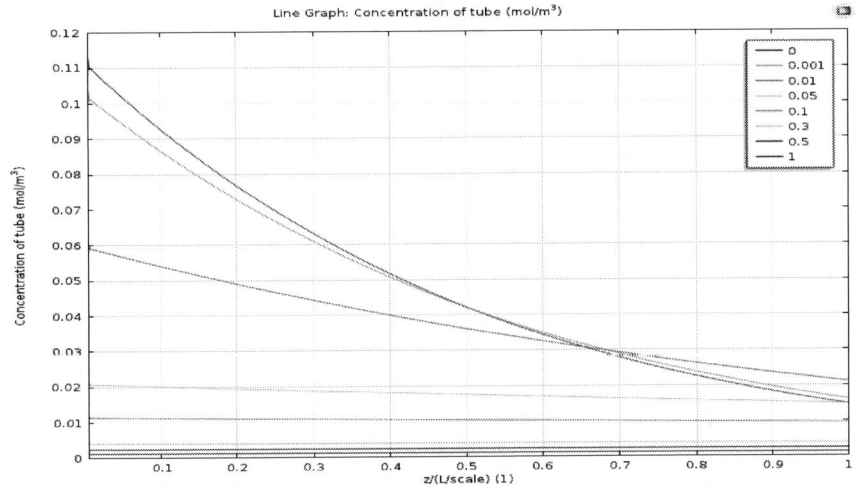

Figure 11. Effect of wettability on carbon dioxide concentrations along the tube.

3.7. The Velocity Distribution in the Tube of the Membrane Contactor

The velocity field at hollow fiber's membrane contactor tube side was simulated by solving Navier-Stokes equations. Figure 12 shows the velocity field at a tube where liquid feed is flowing. Parabolic velocity is associated with the mean velocity along the membrane. As can be seen, at the areas close to the membrane, the velocity reaches a value of about zero sue to non-slip conditions on the membrane walls. Also, the effect of viscosity forces can be seen in this area, which causes to form boundary layer of concentration and velocity in the vicinity of the membrane surface.

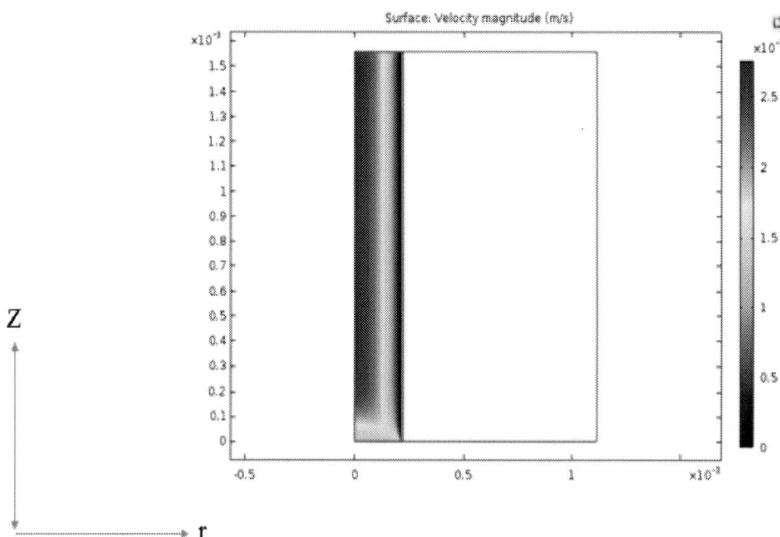

Figure 12. Solvent velocity profile at the tube area from the contactor membrane.

Also, the velocity is along a different axis which shows that in the areas close to the tube inlet, the velocity is not completely developed. This is the main cause of not using a completely developed model and including the input effects which in turn can consider the accuracy and precision of the model with the experimental data. Figure 13 depicts the concentration of carbon dioxide in the radial direction of the liquid phase inside the tube while such a concentration is zero at the feed entering the tube. As can be seen, at the contact surface of membrane-tube (i.e., r = R), carbon dioxide has the highest concentration and consequently the further contact surface, the lower concentration up to zero. This is because it reacts to monoethanolamine and it partly exits along with the liquid through the tube. Also, since the penetration of carbon dioxide to the tube is very low, reaching the tube center looks like an unlimited distance for CO_2. Therefore, the boundary layer is the most important section in the mass transfer process. In doing so, the boundary layer thickness should be decreased to improve the absorption process. This can be done by increasing the liquid phase velocity and thereby enhancing the removal efficiency, as well.

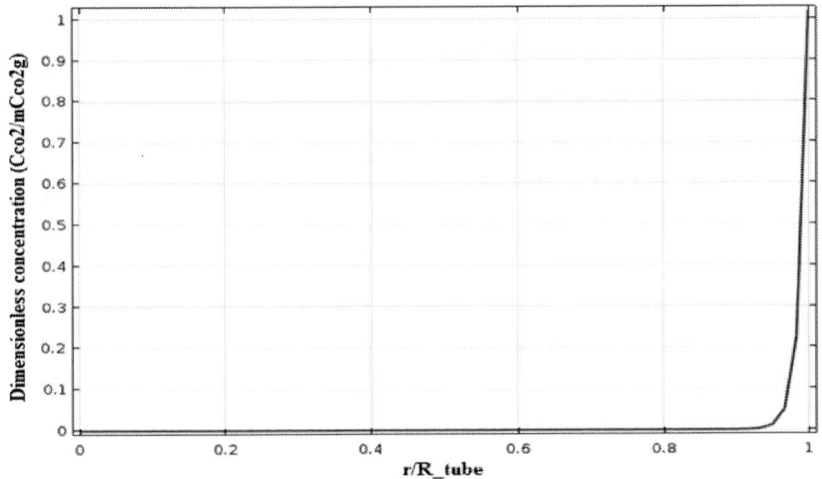

Figure 13. Variations of carbon dioxide concentration at the axial direction in monoethanolamine solution.

3.8. Comparison of CO_2 Removal Efficiency among Concurrent and Counter Flows

With increased concentration gradient along the membrane, the removal efficiency can be enhanced. In doing so, one of the methods to improve the concentration gradient is to use counterflow. In case of counter gas and liquid flows, the concentration gradient will be high along the membrane in both tube and shell sections, improving mass transfer and thus CO_2 removal efficiency. Figure 14 shows CO_2 removal efficiency for both concurrent and counter flows as a function of liquid phase velocity. As can be seen from the figure, with increased liquid phase velocity, the CO_2 removal efficiency is enhanced while such efficiency in counterflow is higher than that of concurrent flow. This is due to the increased concentration gradient along the contactor membrane.

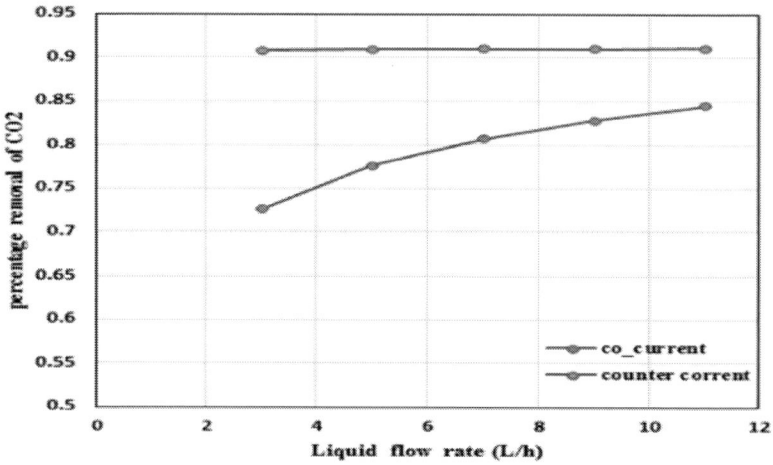

Figure 14. Comparison of carbon dioxide removal efficiency among concurrent and counter flows.

3.9. The Model Validation

Figures 15 and 16 compare the model results to the experimental data. Since mass transfer resistance of CO_2 is from the gas phase to the liquid phase in both gas and liquid phases, increased velocity of each fluid causes to enhance mass transfer from gas to liquid due to decreased mass transfer layer and improved mass transfer factor [30].

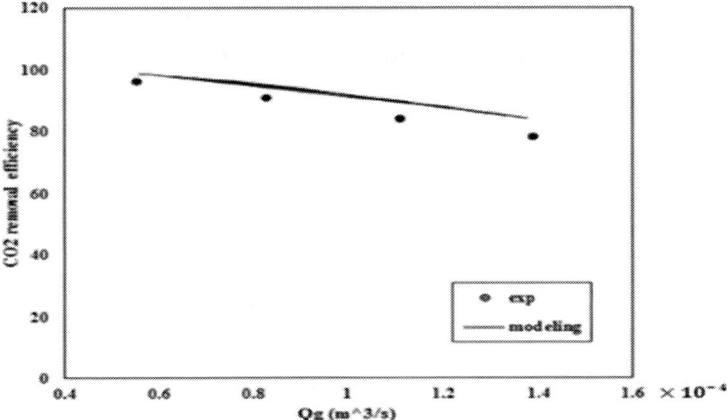

Figure 15. Comparison of the model results to experimental data, carbon dioxide removal efficiency in terms of gas flow rate variations.

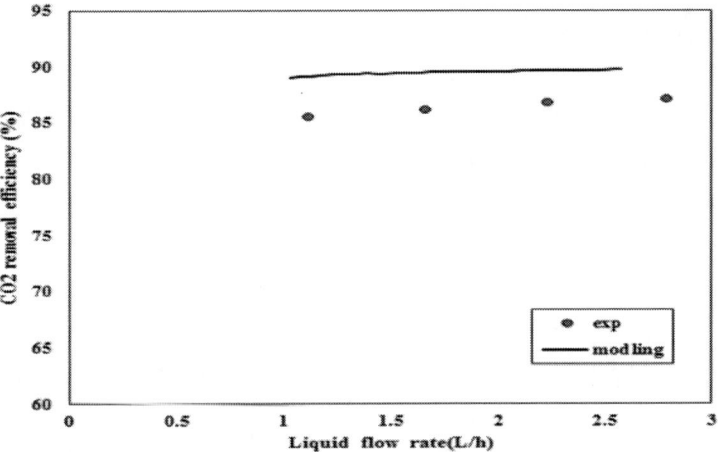

Figure 16. Comparison of the model results to experimental data, carbon dioxide removal efficiency in terms of liquid flow rate variations.

Conclusion

1. The CO_2 concentration is decreased along the tube since it penetrates the tube where the highest concentration of carbon dioxide is in the contact surface of the membrane and absorbent

solution. In doing so, due to the reactions done in the tube between carbon dioxide and monoethanolamine, the CO_2 concentration is decreased and reaches zero at the center of the tube.
2. The lowest mass transfer flux is in the vicinity of membrane and shell contact where fluid velocity is decreased and has a value of zero at the contact surface. Therefore, the molecular influx is mass transfer mechanism with the lowest value since the mass transfer of replacement is decreased and reaches zero at the contactor membrane.
3. With increased volumetric flow rate, CO_2 mass transfer rate to the liquid absorbent is increased since the concentration gradient of carbon dioxide is increased in both gas and liquid phases.
4. The higher liquid phase speed, the higher penetration factor in the liquid phase. In other words, mass transfer resistance is decreased and thus the removal rate is increased. Therefore, output concentration is decreased in the gas phase, as well.
5. With increased gas volumetric flow rate, carbon dioxide removal rate in the contactor will be decreased due to the lower time of stay in the contactor membrane. Therefore, with increasing gas flow rate from 0.6 to 1.4, the CO_2 removal rate will be only 18% decreased. As a result, carbon dioxide absorption does not strongly follow the gas phase velocity.
6. Carbon dioxide removal efficiency strongly follows liquid phase velocity while such a following is stronger in lower velocities and high liquid phase velocity, carbon dioxide absorption process is less dependent on liquid phase velocity.
7. With increased absorbent solvent temperature, carbon dioxide removal efficiency is considerably increased. This is because, with increased absorbent solvent temperature, penetration and velocity of reaction are increased in a solvent which causes to increase carbon dioxide removal range.
8. Wettability has a powerful negative impact on carbon dioxide removal efficiency at the contactor membrane. Consequently, with increased wettability of the membrane, carbon dioxide removal

efficiency is suddenly and considerably decreased while the removal efficiency approaches the zero with increased wettability.
9. With increased wettability, CO_2 concentration is decreased inside the tube since the factor of CO_2 mass transfer to the absorbent solution is lowered and mass transfer resistance is increased.

REFERENCES

[1] Liqin Cao, Xiaohu Wang, Gang Wang, and Jide Wang. *Polymer International*.2015, 64(3), 383–388.

[2] Ajay Kathuria, Mohamad G Abiad, and Rafael Auras. *Polymer International.* 2013, 62(8), 1144–1151.

[3] Khosravi, M. Sadeghi, H. Zare Banadkohi, and M. M. Talakesh. *Ind. Eng. Chem. Res*.2014, 53, 2011−2021.

[4] Pinnau, I.; He, Z. *J. Membr. Sci*.2004, 244, 227−233.

[5] Valeen Rashmi Pereira, Arun M. Isloor, Udaya K. Bhat, A. F. Ismail, Abdulrahman Obaid and Hoong-Kun Fun. *RSC Adv.* 2015,5, 53874-53885.

[6] Moghadam F. Omidkhah M R. Vasheghani-Farahani E. Pedram M Z. Dorosti F. *Separation & Purification Technology*, 201, 77: 128-136.

[7] Bing Yu, Hailin Conga, Xiusong Zhao. *Progress in Natural Science: Materials International*.2012, 22(6), 661–667.

[8] Fletcher, A. J., K. M. Thomas, M. J. Rosseinsky, *J. Solid State Chem.* 178 (2005) 2491–2510.

[9] Paz, F. A. A., J. Klinowski, *Inorg. Chem.* 43 (2004) 3882–3893.

[10] Ge, L., Z. Zhu, F. Li, S. Liu, L. Wang, X. Tang, V. Rudolph, *J. Phys. Chem. C.* 115 (2011) 6661–6670.

[11] Ge, L., Z. Zhu, V. Rudolph, *Sep. Purif. Technol.* 78 (2011) 76–82.

[12] Buonomenna, M. G., W. Yave and G. Golemme.2012,. *RSC Adv.*, 2, 10745-10773

[13] Hudiono, Y. C., T. K. Carlisle, A. L. LaFrate, D. L. Gin, R. D. Noble, *J. Membr. Sci.* 370 (2011) 141–148.

[14] Wijenayake, S. N., N. P. Panapitiya, S. H. Versteeg, C. N. Nguyen, S. Goel, K. J. Balkus Jr., I. H. Musselman, J. P. Ferraris, *Ind. Eng. Chem. Res.* 52 (2013) 6991–7001.
[15] Xue, C., J. Zou, Z. Sun, F. Wang, K. Han, H. Zhu, *Int. J. Hydrogen Energy* 39 (2014) 7931–7939.
[16] Du, H., J. Li, J. Zhang, G. Su, X. Li, Y. Zhao, *J. Phys. Chem. C* 115 (2011) 23261–23266.
[17] Kim, J. H., Y. M. Lee, *J. Membr. Sci.* 193 (2001) 209–225.
[18] Goh, P. S., A. F. Ismail, S. M. Sanip, B. C. Ng, M. Aziz.. *Separation and Purification Technology*.2011, 81 (2011) 243–264.
[19] Soonjae Lee, Jong Suk Lee, Minwoo Lee, Jae-Woo Choi, Sunghyun Kim, Sanghyup Le.2014,. *Journal of Membrane Science*, 452, 311-318.
[20] Ismail, A. F., T. D. Kusworo, A. Mustafa. *Journal of Membrane Science*.2008, 319, 1–2, 306–312.
[21] Shabbir Husain, William J. Koros. *Journal of Membrane Science*.2008, 288, Issues 1–2, 195–207.
[22] Dorosti F. Omidkhah M R. Pedram M Z. Moghadam F. *Chemical Engineering Journal*.2011, 171: 1469-1476.
[23] Suven Pecku, Thilo L. van der Merwe, Heidi Rolfes and Walter W. Focke. *Journal of Vinyl and Additive Technology*.2016, 13, Issue 4, 215–220.
[24] Hassanajili Sh. Masoudi E. Karimi Gh. Khademi M A. *Separation and Purification Technology*.2013, 116: 1–12, 2013.
[25] Kianfar, E., V. Pirouzfar, and H. Sakhaeinia. *J. Taiwan Inst. Chem. Eng.*,2017, **80**, 954.
[26] Salimi, M., V. Pirouzfar, and E. Kianfar . *Colloid Polym. Sci.*,2017, **295**, 215.
[27] Salimi, M., V. Pirouzfar, and E. Kianfar. *Polym. Sci. Ser. A,* 2017,**59**, 566.
[28] Kianfar, E., Salimi, M., Kianfar, F. et al. *Macromolecular Research* 27(1):83–89(2019).
[29] Kianfar, E., Kianfar, F. *Journal of Inorganic and Organometallic Polymers and Materials.* (2019) doi: 10.1007/s10904-019-01177-1.

[30] Wang Z, Fang M, Yu H, Wei CC, Luo Z, *Ind. Eng. Chem. Res.* 2013, 52, 18059-18070.

[31] Mohamed H. Al-Marzouqi, Muftah H. El-Naas, Sayed A. M. Marzouk, Mohamed A. Al-Zarooni, Nadia Abdullatif, Rami Faiz, Modeling of CO_2 absorption in membrane contactors, *Separation and Purification Technology* 59 (2008) 286–293.

In: Computational Fluid Dynamics
Editor: James S. Hutchinson

ISBN: 978-1-53619-756-3
© 2021 Nova Science Publishers, Inc.

Chapter 4

SIMULATION OF HIGH-TEMPERATURE AIR EFFECTS IN HYPERSONIC FLOWS

Yu. Dobrov[1], A. Karpenko[1] and K. Volkov[2,*]
[1]St. Petersburg State University, St. Petersburg, Russia
[2]Faculty of Science, Engineering and Computing,
Kingston University, London, United Kingdom

Abstract

Development and implementation of methods and tools that adequately model fundamental physics and allow credible physics-based optimization for future operational hypersonic vehicle systems are becoming more important due to requirements of ensuring their flight safety. The methods of computational fluid dynamics (CFD) are extensively applied in design and optimization of hypersonic vehicles to get more insight into complex flowfields. Computer simulation is particularly attractive due to its relatively low cost and its ability to deliver data that cannot be measured or observed. Flow discontinuities, high gradients of flow quantities, turbulence effects, flow separation and other flow features impose great demands on the underlying numerical methods. The use of Graphics Processor Units (GPUs) is a cost effective way of improving substantially the performance in CFD applications. GPU platforms make it possible to achieve speedups of an order of magnitude over a standard CPU in many CFD applications. The parallel capabilities of in-house compressible CFD code for hypersonic flow simulations are assessed and

---

*Corresponding Author's Email: k.volkov@kingston.ac.uk.

successful design of a highly parallel computation system based on GPUs is demonstrated. Possibilities of the use of GPUs for the simulation of high-speed and high-temperature flows are discussed. The results obtained are generally in a reasonable agreement with the available experimental and computational data, although some important sensitivities are identified.

Keywords: hypersonic flow, high-temperature air, unstructured mesh, graphics processor unit, parallel algorithm

1. INTRODUCTION

Development and implementation of methods and tools that adequately model fundamental physics and allow credible physics-based optimization for future operational hypersonic vehicle systems are becoming more important due to requirements of ensuring their flight safety [1, 2]. The aero-thermodynamic design and optimization of hypersonic vehicles includes wind tunnel testing, flight experiments and computer simulation [3]. Many aerodynamic and propulsion characteristics still remain uncertain and are difficult to predict due to the lack of flight test data and limitations of ground test facilities [4–6]. The ground-based experimental facilities are not able to reproduce full-scale complex physical and chemical phenomena conditions for flight due to short duration of useful test times [7, 8].

The methods of computational fluid dynamics (CFD) are extensively applied in design and optimization of hypersonic vehicles to get more insight into complex flowfields. Computer simulation is particularly attractive due to its relatively low cost and its ability to deliver data that cannot be measured or observed. Specific challenges involved in modelling and control of hypersonic vehicles, the current state of research and some future directions are described in [9–11]. Flow discontinuities, high gradients of flow quantities, turbulence effects, flow separation and other flow features impose great demands on the underlying numerical methods. The quality of CFD calculations of hypersonic flows strongly depends on the proper prediction of flow physics (non-equilibrium effects, radiation, molecular dissociation, chemical reactions). Investigations of heat transfer, skin friction, secondary flows, flow separation and re-attachment effects demand reliable numerical methods, accurate programming and robust working practices.

Simulation of High-Temperature Air Effects in Hypersonic Flows

At hypersonic speeds, the aero-thermodynamic properties of the air deviate from the perfect gas behavior. High-temperature gas effects alter the aerodynamics of vehicles as their behavior is significantly different to that of a perfect gas. When a vehicle travels through the atmosphere at hypersonic speed, the Mach number is high and the bow shock wave is strong. The temperatures behind the normal shock wave increase with the Mach number. After the air crosses the shock, its enthalpy is sufficient to cause dissociation of air molecules and to induce chemical reactions behind the shock wave [12]. The boundary layer around the vehicle becomes thicker due to high-temperature gas effects of hypersonic flow regime. Thick boundary layer, flow separation and reattachment phenomena lead to viscous behavior in terms of pressure distribution, shock waves and drag, differing significantly from the results of the inviscid flow characteristics [9].

Increasing air temperature above a certain threshold, molecular vibrational mode starts becoming important [3]. The vibrational energy of the molecules becomes excited, and this causes the specific heat capacities to become functions of temperature. Further increase the air temperature leads to molecular dissociation and recombination reactions. As the gas temperature is further increases, chemical reactions occur. For a chemically equilibrium reacting gas specific heat capacities are functions of both temperature and pressure. For higher temperatures, the process of atomic ionization takes place [13]. These processes have a significant impact on the properties of the air. The air can no longer be treated as a perfect gas. For example, at a pressure of 1 atm, oxygen and nitrogen vibrational excitation becomes important above a temperature 800 K. Oxygen dissociation ($O_2 \rightarrow 2O$) begins at about 2000 K, and the molecular oxygen is essentially totally dissociated at 4000 K. At this temperature nitrogen dissociation ($N_2 \rightarrow 2N$) begins, and is totally dissociated at 9000 K. Above a temperature of 9000 K, ions are formed ($O \rightarrow O^+ + e^-$, $N \rightarrow N^+ + e^-$), and the gas becomes a partially ionized plasma. At temperature higher 9000 K, all the nitrogen and oxygen molecules have already dissociated and atomic ionization starts taking place [3].

High-temperature gas effects are a key issue related to hypersonic aerodynamic design and are currently poorly to moderately modelled in CFD [15]. Thin shock layers due to high compression, entropy layers caused by highly swept and curved shock waves, viscous/inviscid interactions and real gas effects are all complex flow features that occur in hypersonic flow [3]. There is also restriction in the speed of computers and a limitation of reliable quantitative data about high-temperature constants [13]. Extensive calculations of the equilibrium properties of high-temperature air, including the equilibrium composition, are presented in [14].

A number of studies have tried to quantify high-temperature gas effects in shock/shock interaction regions using detailed numerical solutions. Significant efforts have focused on simple axisymmetric geometries that are chosen to highlight the important flow field physics in a two-dimensional environment without three-dimensional effects [16–18]. The readiness of CFD to simulate high-enthalpy flows is assessed in [19]. The carbuncle problem represents a local unpleasant displacement of the bow shock wave shape, which strongly affects heat transfer to the vehicle [20]. The analysis of various methods and finite-difference schemes for numerical simulation of reactive flows is performed in [21, 22]. The use of gradient information for the acceleration of uncertainty quantification within the context of viscous hypersonic flows is examined in [23]. Results of simulation of high-enthalpy flows with a Mach number range between 6.4 and 25.9 are provided in [24].

A numerical investigation of vibrational and chemical relaxation of the structure of regular and Mach reflection transition is conducted in [25]. Some degree of vibrational thermal non-equilibrium are modelled using an additional energy equation for the vibrational modes [26]. The vibrational energy source term is modelled using the Landau–Teller model for vibrational relaxation. It is found that real gas effects reduce the angles of the incident and reflected shock waves increasing the angle of the transition from regular to Mach reflection. Dissociation reactions as well as the excitation of vibrational degrees of freedom decrease the Mach stem height. The shock wave/boundary layer interaction causes separation and reattachment regions in flows due to compression [27,28]. High-temperature gas effects nearly double the peak pressure levels in the interaction region and increase the peak heating level [29]. The laminar-turbulent transition in thermally and chemically non-equilibrium boundary layer at hypersonic speeds significantly affects the vehicle performance and surface heating [30]. The shock standoff distance is always employed as the research objective of hypersonic real gas effects or validation purposes of numerical methods [31].

Although a large number of non-equilibrium simulation codes have been developed [32–34], the implementation of the numerical solvers is not considered as fully completed because of limited information with regard to physical model (chemical reactions, transport models) and also validation (lack of both numerical solutions and experimental results for comparison). These methods require extensive computer processor time and storage, and are not generally applicable to parametric and optimization studies or preliminary design calculations. The complex physical phenomena and a wide range of spatial and time scales present in hypersonic flows making the development of efficient and accurate numerical simulation methods challenging.

Speed and accuracy are key factors in the evaluation of CFD solver performance. Advanced simulation tools enable increased spatial and time accuracy, geometric complexity, mesh adaptation, physical process complexity, uncertainty quantification and error control. Unstructured meshes present more flexibility and higher accuracy to represent problems that have complex geometries and boundaries. For enhanced spatial accuracy, high-order algorithms for unstructured meshes are developed. High performance computing (HPC) resources are widely used in aerospace applications. However, the stagnation in the clock-speed of central processing units (CPUs) has led to significant interest in parallel architectures that offer increasing computational power by using many separate processing units. Modern graphics processing units (GPUs) provide new programming models that enable to harness their large processing power and to design CFD simulations at both high performance and low cost [35,36]. The use of GPUs is a cost effective way of improving substantially the performance in CFD applications [37–39]. Achieving good performance for unstructured mesh based CFD solvers on GPUs is more difficult due to their data dependent and irregular memory access patterns [40–42]. GPU platforms make it possible to achieve speedups of an order of magnitude over a standard CPU in many CFD applications.

The motivation of this paper is to assess the in-house compressible CFD code for hypersonic flow simulations and to demonstrate successful design of a highly parallel computation system based on GPUs. The unstructured CFD solver uses an edge-based data structure to give the flexibility to run on meshes composed of a variety of cell types. Possibilities of the use of GPUs for the simulation of high-speed and high-temperature flows are discussed. The results obtained are generally in a reasonable agreement with the available experimental and computational data, although some important sensitivities are identified and discussed.

2. MATHEMATICAL MODEL

Mathematical model used for numerical simulation of hypersonic flows with high-temperature gas effects has to capture the complex thermo-physical phenomena that characterize these flows.

2.1. Governing Equations

A hypersonic vehicle, passing an atmosphere, goes through many different flow regimes due to the change in atmospheric density and pressure with altitude. These regimes are characterized by the Knudsen number. The Boltzmann equation is valid for all flow regimes, from continuum regime to free molecular flow. The Navier–Stokes equations are valid in the continuum regime, below the generally accepted, but often argued, limit of a Knudsen number of 0.01. At lower altitudes where the density is high and the Knudsen number is low, high-speed flows are simulated using traditional CFD techniques based on numerical solution of the Navier–Stokes equations. However, when the Knudsen number becomes larger, the continuum assumption in the Navier–Stokes equations starts to breakdown (Navier–Stokes equations are derived from kinetic theory based on the assumption of small perturbations from an equilibrium velocity distribution function).

The interpretation of physical and chemical phenomena in hypersonic flow applications depends on the assumption of thermal and chemical equilibrium, which allows for a simpler description of the thermal and chemical state of the flow [43]. These processes have characteristic time scales for reaching equilibrium. A single temperature is used to describe the different molecular internal energy modes (thermal equilibrium). This single temperature describes the energy modes of all molecules and it is the same as the temperature of the surrounding. All chemical reactions are in balance, and the system does not spontaneously undergo any change in chemical composition (chemical equilibrium). When the characteristic flow time is much shorter than the time to complete chemical reactions or energy exchange mechanisms, then the flow is in a non-equilibrium state [3].

The shock angle decreases if Mach number increases. This makes the shock layer thinner. At the same time, the thickness of the boundary layers on hypersonic vehicle is thicker in comparison to the thickness of the shock layer. The laminar boundary layer thickness on a flat plate grows as $\delta \sim M^2/Re_x^{1/2}$, where M is the free stream Mach number and Re_x is the local Reynolds number. For large Mach numbers, the boundary layer occupies a significant portion of the flowfield. In some cases, it merges with the bow shock itself. Therefore, viscous and inviscid effects interact strongly in hypersonic flows.

The unsteady three-dimensional flow of the viscous compressible gas is described with a system of equations including mass conservation equation, momentum conservation equation and energy conservation equation. Integration over control volume V allows to present governing equations in the integral form

$$\frac{\partial}{\partial t} \oint_V \boldsymbol{U} dV + \oint_{\partial V} \boldsymbol{F} \cdot d\boldsymbol{S} = 0, \qquad (1)$$

where $\boldsymbol{U}$ is the vector of conservative variables in point x at time t, $\boldsymbol{F}$ is the flux tensor with dimension 3×5, $d\boldsymbol{S} = \boldsymbol{n} dS$ is the surface area vector, $\boldsymbol{n}$ is the surface normal vector. The vector of conservative variables and flux tensor are presented in the form

$$\boldsymbol{U} = \begin{pmatrix} \rho \\ \rho \boldsymbol{v} \\ \rho e \end{pmatrix}, \quad \boldsymbol{F} = \begin{pmatrix} \rho \boldsymbol{v} \\ \rho \boldsymbol{v}^* \otimes \boldsymbol{v} + p\boldsymbol{I} - \boldsymbol{\tau} \\ (\rho e + p)\boldsymbol{v} - \boldsymbol{v} \cdot \boldsymbol{\tau} + \boldsymbol{q} \end{pmatrix}.$$

The viscous terms in the governing equations (viscous stresses and heat fluxes) require constitutive relations to relate these viscous terms to the flow and thermodynamic variables. The components of the viscous stress tensor have the form

$$\tau_{ij} = \mu \left(\frac{\partial v_i}{\partial x_j} + \frac{\partial v_j}{\partial x_i} - \frac{2}{3} \frac{\partial v_k}{\partial x_k} \delta_{ij} \right).$$

The heat flux vector is found from the Fourier law

$$\boldsymbol{q} = -\lambda \nabla T,$$

Here, t is the time, ρ is the density, p is the pressure, T is the temperature, $\boldsymbol{v}$ is the velocity vector with components v_x, v_y and v_z in cartesian directions x, y and z, e is the total energy per unit mass, $\boldsymbol{q}$ is the heat flux, $\boldsymbol{\tau}$ is the viscous stress tensor, $\boldsymbol{I}$ is the identity tensor. Star denotes the conjugate tensor.

The equation of state of a perfect gas has the form

$$p = \rho \frac{R_0}{M(p,T)} T,$$

where R_0 is the universal gas constant, M is the molecular weight of the mixture, which depends on the pressure and temperature.

The total energy is equal to the sum of the internal energy (it includes the energies of translational, rotational, vibrational motions and electronic excitation of atomic and molecular components of the gas mixture) and kinetic energy

$$e = \varepsilon + \frac{1}{2}|v|^2,$$

where ε is the specific internal energy. It is assumed that there is equilibrium over all internal degrees of freedom of particles with an energy of translational motion. Using the enthalpy $h = \varepsilon + p/\rho$, the expression for the total energy is written as

$$e = h + \frac{1}{2}|v|^2 - \frac{p}{\rho}.$$

Specific enthalpy of the gas is

$$h = h_0 + \int_{T_0}^{T} c_p dT,$$

where h_0 is the enthalpy at the reference temperature T_0. For a perfect gas, the molecular weight and the gas constant ($R = R_0/M$), the specific heat capacities at constant pressure and constant volume are constant and the Mayer's relationship takes place ($c_p - c_v = R$).

The Sutherland's law is used to obtain molecular viscosity as a function of temperature

$$\frac{\mu}{\mu_*} = \left(\frac{T}{T_*}\right)^{3/2} \frac{T_* + S_0}{T + S_0},$$

where $\mu_* = 1.68 \times 10^{-5}$ kg/(m·s) is the viscosity at the reference temperature $T_* = 273$ K, and $S_0 = 110.5$ K for air. Sutherland's law is accurate for air over a range of several thousand degrees and is appropriate for hypersonic viscous flow calculations. The thermal conductivity is obtained from dynamic viscosity and the Prandtl number, $\lambda = c_p \mu / \text{Pr}$. For air at standard conditions, $\text{Pr} = 0.71$.

2.2. Chemical Reactions

Vibrational and chemical processes take place as a result of molecular collisions. The actual number of collisions depends on the type of molecule and the relative kinetic energy between the two colliding particles. In turn, molecular collisions take time to occur. The precise interval of time depends on the molecular collision frequency $Z \sim p/T^{1/2}$. The collision frequency is low for low pressures and high temperatures. The equilibrium system assumes that the gas has enough time for the necessary collisions to occur and the properties of the system at a fixed pressure and temperature are independent on time (reactions take place at equal rates in their direct and reverse directions).

The chemical equilibrium assumption consists of considering that all the species in the system have enough time to reach the equilibrium state. In general, if the mixture has N_s species and N_e elements, then $N_s - N_e$ independent chemical equations with the appropriate equilibrium constants are considered. The remaining equations are obtained from the mass balance equation and Dalton's law of partial pressures.

In numerical calculations, air is represented as a mixture of $N_s = 11$ different ideal gases (e^-, N, O, Ar, N_2, O_2, NO, N^+, O^+, Ar^+, NO^+). The distribution of chemical elements is given by the molar fraction $\chi_i^0 = \nu_i^0/\nu_\Sigma^0$, where $\nu_\Sigma^0 = \sum_i^{N_e} \nu_i^0$, and ν_i^0 is the number of moles of the chemical element i. The superscript 0 indicates that the number of moles is considered for the chemical element, and not for the molecule or component of the mixture. The partial pressure of an ideal gas in a mixture equals the pressure that is a gas if it occupies the same volume as the entire gas mixture at the same temperature. The partial pressure of the gas mixture component k is related to the mole fractions by the relation $\chi_k = \nu_k/\nu_\Sigma = p_k/p_\Sigma$, where p_Σ is the pressure of the mixture. To simplify the formulation of the equilibrium equations, it is assumed that upon dissociation of any molecule, its dissociation occurs up to the atoms of the elements. Therefore, the number of reactions is equal to the difference of accounted components of the mixture N_s and the number of considered chemical elements N_e ($N_l = N_s - N_e$, where $l = 1, \ldots, N_l$).

The system of equations describing the equilibrium composition of the mixture includes the following equations

$$\sum_{k}^{N_s} \phi_i^k \chi_k = \chi_i^0 \frac{\nu_\Sigma^0}{\nu_\Sigma} \quad (i = 1, \ldots, N_e);$$

$$p_\Sigma^{\beta^l} \prod_{k}^{N_s} \chi_k^{\eta_k^l} = K_p^l, \quad \beta^l = \sum_{k}^{N_s} \eta_k^l \quad (l = 1, \ldots, N_l); \qquad (2)$$

$$\sum_{k}^{N_s} \chi_k = 1.$$

Here, ϕ_i^k is the number of particles of the chemical element i in the mixture component k, $\chi_k = \nu_k/\nu_\Sigma$ is the molar fraction of gas mixture components, ν_k is the number of moles of the component of the mixture, $\nu_\Sigma = \sum_k^{N_s} \nu_k$ is the total number of moles of the gas mixture in the current state, K_p^l is the constant of chemical equilibrium with respect to pressure (subscript p) of the reaction l, η_k^l are the stoichiometric coefficients for the reaction l, which are negative for the initial products and positive for the reaction products. In the system (2), there are $N_s + 1$ equations with unknowns ν_Σ^0/ν_Σ and $\chi_1, \chi_2, \ldots, \chi_{N_s}$.

The equilibrium constant K_p^l is related to the Gibbs energy for products participating in this reaction by the relation

$$RT \ln K_p^l = -\sum_{k}^{N_s} \eta_k^l \Delta G_k.$$

The change in the Gibbs energy of the component of the mixture k from the standard state to the state at the given temperature, ΔG_k, is expressed as

$$\Delta G_k(T) = \Delta H_k(T) - T\Delta S_k(T),$$

where $\Delta H_k(T)$ is the change in molar enthalpy from the standard state, $\Delta S_k(T)$ is the change in entropy. The functions $\Delta H_k(T)$ and $\Delta S_k(T)$ for each component of the mixture k are calculated using polynomial approximations [46]. The equilibrium constant is found from the equation

$$R \ln K_p^l = \sum_{k}^{N_s} \eta_k^l \Delta \Phi_k(T) - \frac{1}{T} \sum_{k}^{N_s} \eta_k^l \Delta H_k(0).$$

The change in the Gibbs energy, $\Delta G_k(T)$, is expressed in terms of the reduced Gibbs function, $\Delta \Phi_k(T)$, and the formation enthalpy, $\Delta H_k(0)$.

2.3. High-Temperature Effects

In the chemically equilibrium reacting flows, the chemical components computation is time consuming. The explicit expressions derived in [47] for equilibrium air are based on density and internal energy as the independent variables. These expressions are used to determine the other thermodynamic and transport properties, avoiding calculating chemical species. When high-temperature effects are taken into account, the system of equations (2) is not solved, and explicit expressions for the molecular weight and enthalpy as functions of pressure and temperature, $M_\Sigma = M_\Sigma(p, T)$ and $h = h(p, T)$, proposed in [47], are applied to simplify calculations.

The model proposed in [47] takes into account the dissociation of oxygen and nitrogen, formation of nitric oxide, possibility of the appearance of an electronic component due to the single and double ionization of oxygen, nitrogen and argon. The oxygen dissociation reaction takes place, followed by reactions of nitrogen dissociation and single ionization of a weighted average mixture of these gases. When considering each subsequent reaction, the previous one is considered as fully completed. In the temperature range from 200 to 20000 K and pressures from 0.001 to 1000 atm, the error of the model does not exceed 1.5% by density and 3% by enthalpy.

To extend the existing numerical methods developed for a perfect gas to hypersonic flows with high-temperature effects, an effective ratio of specific heat capacities is introduced. It makes it possible to calculate the speed of sound from the relation

$$a^2 = \gamma_* \frac{p}{\rho}.$$

The relation between the effective ratio of specific heat capacities, γ^*, and the ratio of specific heat capacities of perfect gas, $\gamma = c_p/c_v$, is

$$\gamma_* = \gamma \frac{\rho}{p \rho_p}.$$

For isentropic conditions ($ds = 0$), the speed of sound is calculated as

$$a^2 = \left(\frac{\partial p}{\partial \rho}\right)_s = \left(\rho_p + \rho_T \frac{dT}{dp}\right)^{-1}.$$

In thermodynamic equilibrium, the relation $T ds = dh - dp/\rho$ is valid. Since $dh = h_p dp + h_T dT$, then

$$\frac{dT}{dp} = \frac{1 - \rho h_p}{\rho h_T}.$$

The specific heat capacity at constant pressure is calculated using central finite differences of the second order

$$c_p(p, T) = \left(\frac{\partial h}{\partial T}\right)_p = \frac{h(p, T + \Delta T) - h(p, T - \Delta T)}{2\Delta T},$$

where $\Delta T = 0.01T$.

For shock wave calculations, the effective adiabatic index is defined as $\gamma_s = h/\varepsilon$, and the specific internal energy is found from the relation $\varepsilon = p/[(\gamma_s - 1)\rho]$. The effective ratio of specific heat capacities is defined as

$$\gamma_e = \frac{c_p}{c_v} \left[1 + \frac{p}{M} \left(\frac{\partial M}{\partial p}\right)_T\right]^{-1},$$

where M is the molar mass.

3. COMPUTATIONAL PROCEDURE

The governing equations are solved with in-house finite volume code run on unstructured meshes consisting of cells of various topology.

3.1. Numerical Method

The non-linear solver works in an explicit time-marching fashion, based on a Runge–Kutta stepping procedure. The unstructured CFD code uses an edge-based data structure to give the flexibility to run on meshes composed of a variety of cell types. The fluxes through the surface of a cell are calculated on the basis of flow variables at nodes at either end of an edge, and an area associated with that edge (edge weight). The edge weights are pre-computed and take into account geometry of the cell. The flux vector is split into the inviscid and viscous components. The governing equations are solved with upwind finite difference scheme for inviscid fluxes, and central difference scheme of the second order for viscous fluxes [48]. For simulation of low-speed flows, convergence to a steady state is accelerated by the use of low-Mach number preconditioning method. The computational procedure involves reconstruction of the solution in each control volume and extrapolation of the unknowns to find the flow variables on the faces of control volume, solution of Riemann problem for each face of the control volume [49], and evolution of the time step.

The governing equation (1) is written in the form

$$\frac{dU_i^n}{dt} + L(U_i^n) = 0, \qquad (3)$$

where $L(U_i^n)$ is the spatial differential operator. The subscript i refers to the control volume, and the superscript n refers to the time layer. The flow variables vector averaged over the control volume V_i is

$$U_i = \frac{1}{V_i} \int_{V_i} U \, dV.$$

The flow residual within each cell of unstructured mesh is given by the sum of the normal inviscid and viscous fluxes over all faces

$$R_i(U) = -\frac{1}{V_i} \sum_j^{N_i} F_{ij}(U) S_{ij}, \qquad (4)$$

where F_{ij} is the numerical flux from cell i through the face j at the face center, V_i is the volume of cell i, S_{ij} is the area of the face j of the cell i, N_i is the number of faces of the cell i.

The inviscid flux is calculated using gradient reconstruction of primitive variables. The calculations of gradients are based on Green–Gauss theorem. The inviscid flux is found from the relation

$$F(U_L, U_R) = \frac{1}{2}\Big[F(U_L) + F(U_R) - |A|(U_R - U_L)\Big].$$

where the subscripts L and R refer to cells on the left and on the right edges of the control volume. The matrix A is presented in the form $A = R\Lambda L$, where Λ is the diagonal matrix composed from the Jacobian eigenvalues, and R and L are the matrices composed from its right and left eigenvectors, respectively. The flow variables are found as

$$U'_L = U_L + \varphi_L \nabla U_L \cdot \Delta r_L;$$
$$U'_R = U_R + \varphi_R \nabla U_R \cdot \Delta r_R.$$

Here, Δr_L and Δr_R are the distance vectors from the cell center to the face center (Figure 1a), ∇U_L and ∇U_R are the gradients calculated in the cell center, φ_L and φ_R are the limiter functions in the cell. Gradients of flow variables are found from expression

$$(\nabla U)_i^l = \frac{1}{V_i} \sum_{k=1}^{N_i} \frac{1}{2} (U_i + U_k) n_l S_{ik},$$

where n_l is the component of normal vector. Wall boundary conditions are realized using ghost cells (Figure 1b).

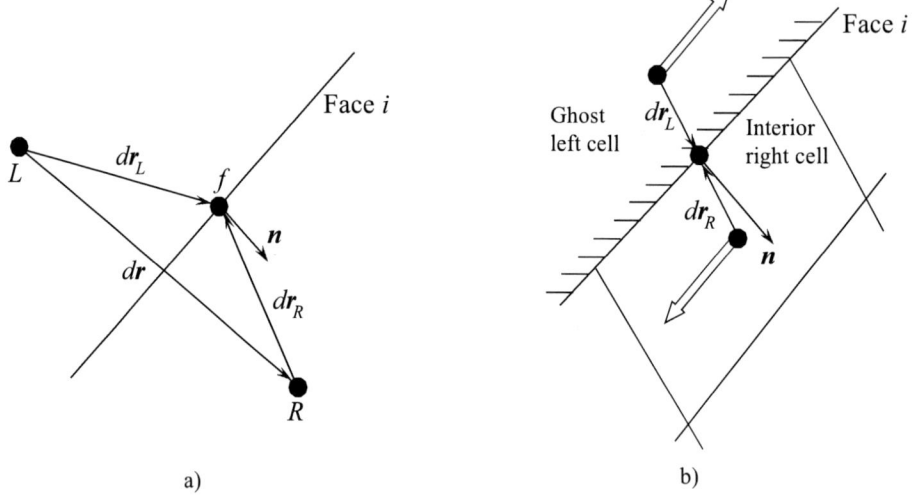

Figure 1. Face of control volume (a) and wall boundary conditions (b).

3.2. Chemical Kinetics

The calculation of a chemical equilibrium composition is conceptually straightforward. New variables $x_1 = \ln \chi_1$, $x_2 = \ln \chi_2$, ..., $x_{N_s} = \ln \chi_{N_s}$, $x_{N_s+1} = \nu_\Sigma^0 / \nu_\Sigma$ are introduced to solve the system of equations (2). Taking the logarithm of the equations of chemical equilibrium gives

$$\sum_{k}^{N_s} \phi_i^k e^{x_k} - \chi_i^0 x_{N_s+1} = 0 \quad (i = 1, \ldots, N_e);$$

$$\sum_{k}^{N_s} \eta_k^l x_k + \beta^l \ln p_\Sigma - \ln K_p^l = 0, \quad \beta^l = \sum_{k}^{N_s} \eta_k^l \quad (l = 1, \ldots, N_l); \quad (5)$$

$$\sum_{k}^{N_s} e^{x_k} - 1 = 0.$$

The linearization of the equations (5) using Taylor series expansion leads to the equations

$$\sum_{k}^{N_s} \phi_i^k \chi_k \Delta x_k - \chi_i^0 \Delta x_{N_s+1} = -\left(\sum_{k}^{N_s} \phi_i^k \chi_k - \chi_i^0 x_{N_s+1} \right) \quad (i = 1, \ldots, N_e);$$

$$\sum_{k}^{N_s} \eta_k^l \Delta x_k = -\left(\sum_{k}^{N_s} \eta_k^l x_k + \beta^l \ln p_\Sigma - \ln K_p^l \right) \quad (l = 1, \ldots, N_l);$$

$$\sum_{k}^{N_s} \chi_k \Delta x_k = -\left(\sum_{k}^{N_s} \chi_k - 1 \right).$$

The linearized system is written with respect to unknown increments in the matrix form

$$A \Delta x = b,$$

where A is the matrix of coefficients, b is a vector of right-hand sides, and Δx is a change in the solution vector between the iterations. The solution at the iteration $n + 1$ is found as

$$x^{n+1} = x^n + (1 + \theta) \Delta x,$$

where θ is the coefficient of lower relaxation, which is used to ensure the stability of the numerical procedure.

To solve the system of equations (2), it is necessary to set the total pressure of the mixture p_Σ and the temperature T. The chemical composition of the reacting mixture is expressed in terms of mole fractions, χ_k. The molecular weight and specific enthalpy of the mixture are found from the relationships

$$M_\Sigma = \sum_{k}^{N_s} M_k \chi_k, \quad h = \frac{1}{M_\Sigma} \sum_{k}^{N_s} H_k \chi_k,$$

where M_k is the molecular weight of the mixture component k.

4. PARALLEL ALGORITHM

The computational procedure is implemented as a computer code in C/C++ programming language. Parallelization of the computational procedure is performed by a message passing interface (MPI). CUDA technology is used to implement GPU version of the solver. An equivalent solver is made in C++ to be run on CPU for benchmarking purposes.

The computation steps required by the problem considered are classified into two groups, computations associated with faces and edges, and computations associated with volumes. The numerical scheme exhibits a high degree of data parallelism because the computations at each edge/volume are independent with respect to the computations performed at the rest of edges/volumes. Moreover, the explicit scheme presents a high arithmetic intensity and the computation exhibits a high degree of locality.

Explicit time-marching algorithms are the most convenient ones to be ported on to the GPU. This is because there is no iteration, and the new value of a variable depends only on the old time values. Hence, the update of a given variable is done independent of variables being updated on other threads. There is no recursive relation between the variables on the threads, since they are all known at the old time step. However, even for explicit algorithms, a few changes are needed for efficiently implementation of numerical algorithms on the GPU. These are related to the use of shared memory and the layout of data structures. Memory coalescing and block size influence the speed achieved. The data should be organized such that adjacent threads access adjacent nodal data. In addition, data should be, if it is possible, copied to shared memory and reused as much as possible. Therefore, even explicit algorithm based CFD codes need to be reorganized to take advantage of the GPU architecture [41,42].

The implementation is split between CPU and GPU. Pre- and postprocessing steps are done on the CPU, leaving only the computation itself to be performed on the GPU. For example, the CPU constructs the mesh and evaluates the face areas, face normals and cell volumes. The initialization of the flowfield is also done on the CPU. Each time step of the computation then involves a series of kernels on the GPU which evaluate the cell face fluxes, sum the fluxes into the cell, calculate the change in properties at each node, smooth the variables and apply boundary conditions. Each kernel operates on all the nodes (no distinction is made between boundary nodes and interior nodes). This causes difficulties if an efficient code is to be obtained. For example, the change in a flow property at a node is formed by averaging the flux sums of the adjacent cells (for mesh with quadrangle cells, four cells surround an interior node, but only two at a boundary node). The indices of the cells required to update a node are pre-computed on the CPU and loaded into GPU texture memory. For a given node, the kernel obtains the indices required and then looks up the relevant flux sums which are stored in a separate GPU texture. This avoids branching within the kernel.

A block-scheme of the parallel computational algorithm is shown in the Figure 2. The main stages are identified and the main sources of data parallelism are represented indicating that the calculation affected by it are performed simultaneously for each data item of a set (the data items represent the volumes or faces/edges of the finite volume mesh). Time stepping process is repeated until the final simulation time is reached. At the $(n+1)$-th time step, the residual is evaluated to update the state of each cell. In order to add the contributions associated with each edge, two variables are used in the algorithm for each volume. The first variable is used to store the contributions to the local time step size of the volume, and the second variable is used to store the sum of the contributions to the state of cell.

The most costly stage in the algorithm is edge-based calculations involving two calculations for each face communicating two cells. This contribution is computed independently for each face and is added to the partial sums associated to each cell. For each control volume, the local time step is computed. The computation for each volume does not depend on the computation for the rest of volumes and therefore this stage is performed in parallel. The minimum of all the local time steps previously obtained for each volume is computed. The $(n+1)$-th state of each control volume is approximated from the n-th state using the data computed in the previous phases. This stage is also completed in parallel.

The implementation of the finite volume method using a global memory and register file is illustrated in the Figure 3. Each time layer calculations are performed in two stages. Two kernels are used for the parallel implementation of the finite volume method on GPU, one of which calculates the flow through the faces of control volumes (stage 1, Figure 3a), and the other one provides flow variable calculations on the next time layer (stage 2, Figure 3b). On the first stage, flow variables in the centers of control volumes are stored in global

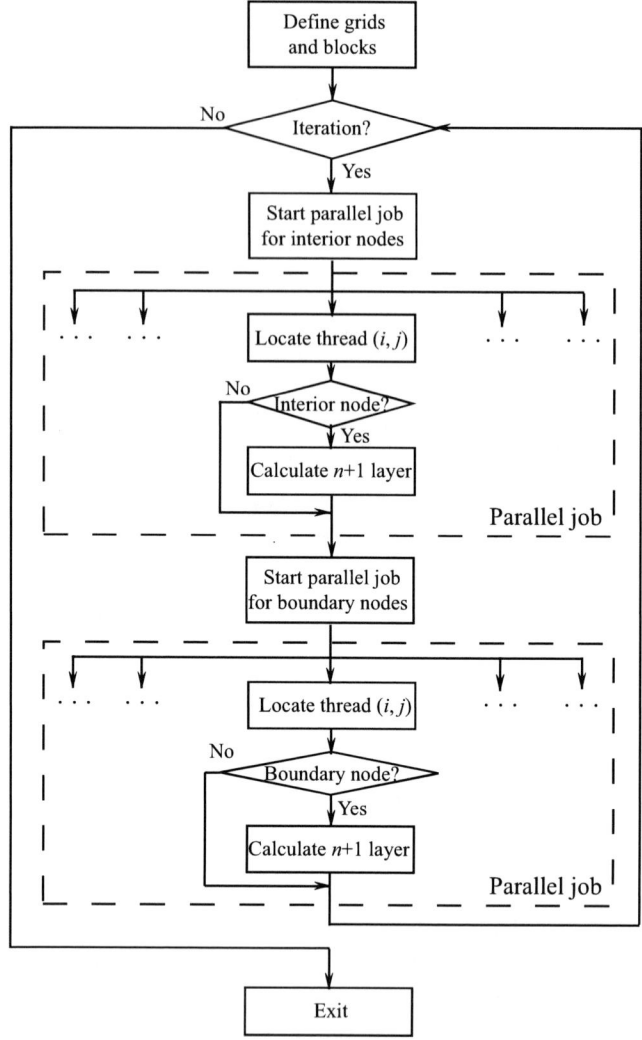

Figure 2. Main stages of the parallel algorithm.

memory. One thread is used to calculate the fluxes through the faces of control volume. Each thread uses the flow variables vector in adjacent control volumes, i and $i+1$. On the second stage, a set of threads corresponding to the same number of control volumes is launched to calculate the flow variables vector on a new time level. The fluxes through the faces $i - 1/2$ and $i + 1/2$ are used, and the solution is computed in the control volume i.

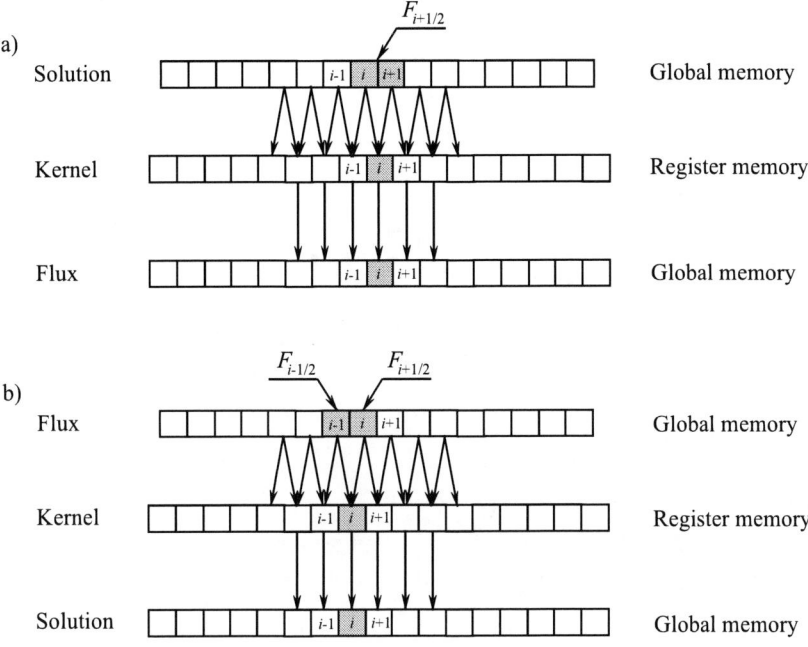

Figure 3. Flux calculation (*a*) and calculation of flow variables on a new time layer (*b*).

To compare the performance of the CFD solver on CPU and GPU, similar algorithms for calculating the perfect gas and equilibrium air are implemented on CPU. Calculations are made on two different servers. Server 1 uses one core of the Xeon E5-2680 v3 CPU or one NVidia Tesla K40 calculation module. Server 2 uses one IBM Power8 CPU thread or one NVidia Tesla K40 module. Some characteristics of the servers used in CFD calculations are presented in the Table 1.

Table 1. Parameters of GPU hardware

Server	Server 1		Server 2	
Brand name	Huawei RH2288H		IBM S822LC	
RAM per node, GB	128		256	
Number of CPUs	2		2	
Number of GPUs	2		2	
Processor unit	Xeon E5-2680 v3	Tesla K40	IBM Power8	Tesla P100
Production year	2014	2013	2016	2016
Frequency (base), MHz	2500	745	2860	1328
Number of cores	12	2880	10	3584
Peak performance, GFlops (single precision, base)	960	4291	458 458	9519
Peak performance, GFlops (double precision, base)	480	1430	229	4760
Memory bus width, bit	64	384	64	4096
RAM per socket, Gb	768	12	512	16
Bandwidth of memory bus per socket, Gb/s	68	288	115.2	732

5. COMPOSITION OF AIR

The numerical calculations are carried out for the model of air consisting of 11 species (e^-, N, O, Ar, N_2, O_2, NO, N^+, O^+, Ar^+, NO^+). The air represents the mixture of 78% N_2, 21% O_2 and 1% Ar.

Composition of high-temperature air is shown in the Figure 4 and Figure 5 where the equilibrium composition of high-temperature air (in terms of mole fraction) is given as a function of temperature at $p = 0.01$ atm. The oxygen begins to dissociate above 2100 K and is completely dissociated above 7800 K. The nitrogen begins to dissociate above 6000 K and is completely dissociated above 12000 K. The NO is present between 800 and 11000 K, with a peak mole fraction occurring about 2500 K. The line for O has a local maximum around 4000 K and then decreases at higher temperatures in the range from 6000 to

20000 K. This does not mean, however, that the total amount of O atoms is decreasing in this range. Rather, it is a consequence of the definition of mole fraction. Because $O = \sum O / \sum$, where $\sum O$ is the number moles of O and $\sum$ is the total number of moles, then O decreases simply because the total number of moles of the mixture is increasing (as a result of the dissociation of N_2, for example). The data plotted are in a good agreement with the detailed tabulations found in [14].

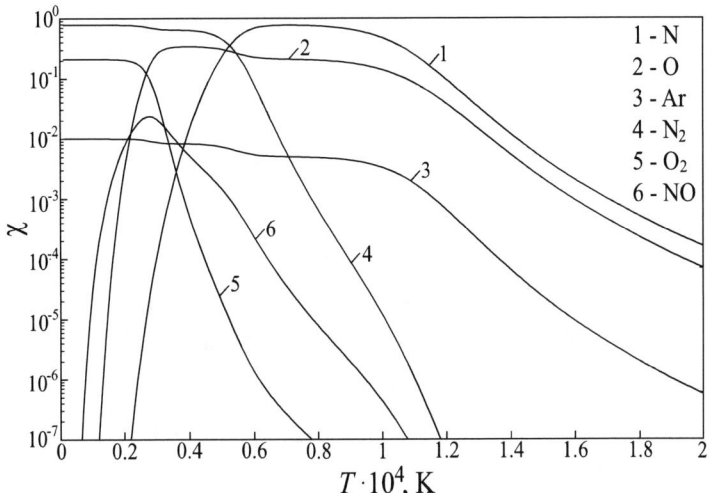

Figure 4. Composition of equilibrium air as a function of temperature at pressure 0.01 atm ($1 — N, 2 — O, 3 — Ar, 4 — N_2, 5 — O_2, 6 — NO$).

If the pressure is increased to 10 atm then all of the lines would qualitatively shift to the right, that is, the various dissociation processes would be delayed to higher temperatures. On the other hand, if pressure is decreased, then all of the lines would qualitatively shift to the left, that is, dissociation would occur at lower temperature. Hence, raising the pressure decreases the amount of dissociation, and lowering the pressure increases the amount of dissociation.

The equilibrium speed of sound is a function of both pressure and temperature, unlike the case for a calorically or thermally perfect gas where it depends on temperature only [3]. This is emphasized in the Figure 6 (effective specific heat capacity ratios, $\gamma_c = c_p/c_v$, γ_s and γ_e are shown), which gives the equi-

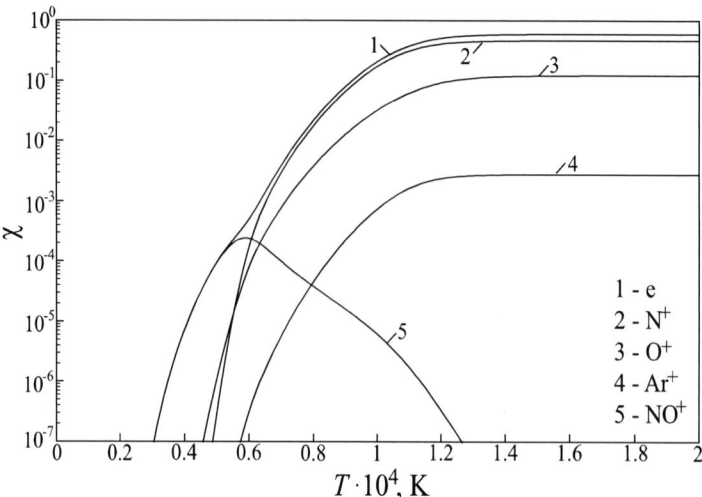

Figure 5. Composition of equilibrium air as a function of temperature at pressure 0.01 atm ($1 - e$, $2 - N^+$, $3 - O^+$, $4 - Ar^+$, $5 - NO^+$).

librium speed of sound for high-temperature air as a function of both pressure and temperature. The difference between the frozen and equilibrium speeds of sound in air can be as large as 20% under practical conditions [43]. In turn, this once again underscores the ambiguity in the definition of Mach number for high-temperature flows.

6. RESULTS AND DISCUSSION

The GPU version of the CFD code is validated for a variety of benchmark test cases. The benchmark problems provide insight into the ability of CFD solver and numerical schemes to capture the bow shock, smoothly resolve the post-shock stagnation region flow and predict a smoothly varying heating distribution around the stagnation point.

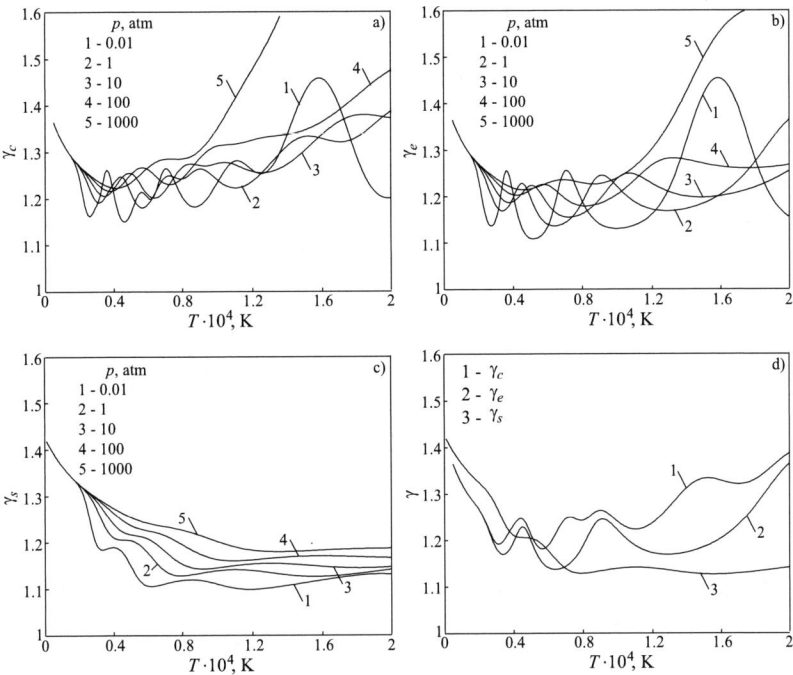

Figure 6. The effective ratios of specific heat capacities for air as functions of pressure and temperature (a, b and c) and the effective ratios of specific heat capacities for air as functions of temperature at the fixed pressure ($p = 1$ atm).

6.1. Shock Tube Problem

The shock tube test case considers a long tube ($L = 10$ m) containing a gas separated by a thin diagram located at $x = 4$ m. The gas is assumed to be at rest on both sides of the diagram ($u_L = u_R = 0$), but it has different constant pressures ($p_L = 10^5$ Pa and $p_R = 10^4$ Pa) and densities ($\rho_L = 1.0$ kg/m^3 and $\rho_R = 0.125$ kg/m^3) on each side. The mixture of perfect gases consists of 0.7778 N$_2$ and 0.2222 O$_2$ (by mass fraction). The chosen composition is closely match the analytical model for air at the given temperatures. It is a test of the perfect gas mixture if it can reproduce the solution computed analytically. At time $t = 0$, the diagram is removed, and the problem is to determine ensuing flow of the gas. The solution of this problem consists of a shock wave moving into the low pressure region, a rarefaction wave that expands into the high pressure region, and a contact discontinuity which represents the interface.

Unstructured tetrahedral mesh is applied to solve 3D shock tube problem. Calculations are based on meshes with a different number of cells. The coarsest mesh contains about 10^4 cells (mesh 1), and the finest mesh contains about 10^7 cells (mesh 4). The intermediate meshes contain 10^5 cells (mesh 2) and 10^6 (mesh 3) cells. The time step is 1.52×10^{-5} s, and the total computational time is 7.63×10^{-3} s. Courant number is equal to 0.85. The calculations are performed on one module of Tesla S1070 platform with 1.44 GHz (a number of cores is 256), and one core of CPU AMD Phenom 2 with 3 GHz.

The numerical results, shown in the Figure 7, indicate higher resolved solutions for a given time step and given mesh size than the analytical solution. The results computed numerically do not have any non-physical oscillations on shock or contact discontinuities. The computed solution shows a good agreement with the analytical one. This is a fair indication that the model based on the mixture of perfect gases has been implemented correctly. The discrepancies near the shock are the standard numerical errors involved with any finite volume solution. Most of the discrepancies occur in the regions where the flow changes most rapidly. This behavior is expected in any scheme where discontinuities are captured over a few cells that have large flow quantity gradients.

Time of calculation of 1000 time steps and speedup of calculations are presented in the Table 2 (time is given in seconds). Two indices are used to specify computational option. The first index corresponds to the time-marching scheme used in calculations based on one-step (option A) or three-step (option B) Runge–Kutta time-stepping procedure. The second index corresponds to the exact Godunov (index 1) or approximate Roe (index 2) Riemann solvers.

6.2. Hypersonic Flow around Sphere

Accurate simulation of stagnation region heating in hypersonic flows is a key requirement for acceptance of any algorithm proposed for use in aerothermodynamic analysis [43]. The shock standoff distance on sphere in hypersonic flow is one of the most appropriate parameter for validation of CFD results. For a high Mach number, the shock standoff distance is much smaller than the sphere radius. Its experimental measurement is difficult, and large errors have to be accepted. Some significant theoretical studies have been performed to predict the shock standoff distance on spheres in hypersonic flows [50, 51].

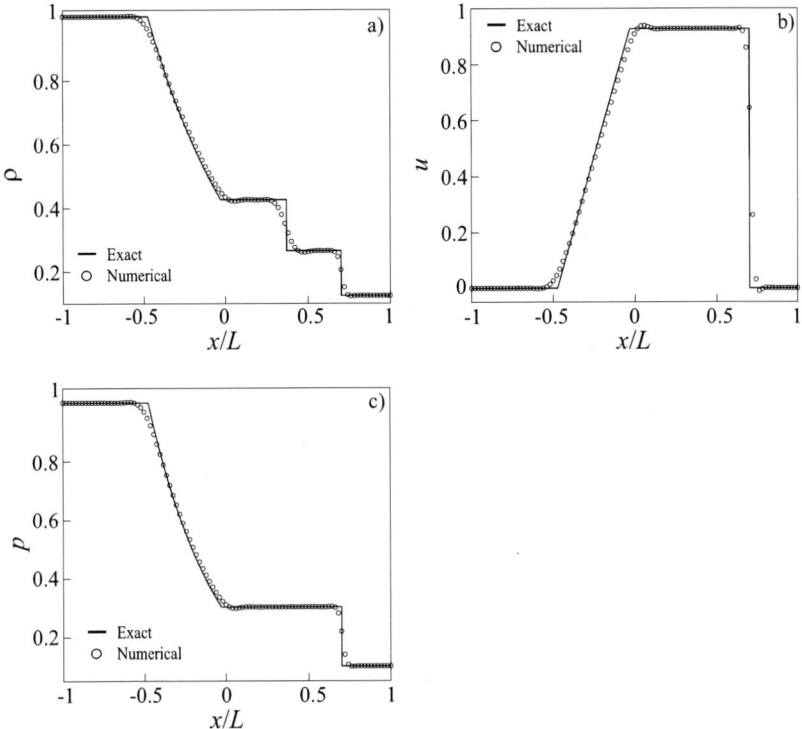

Figure 7. Solution of shock tube problem: a) density, b) velocity, c) pressure.

A theoretical model to find the shock standoff distance on axisymmetric blunt bodies for frozen and equilibrium flows is developed in [52]. The shock standoff distance becomes larger as the flow condition is changed from equilibrium to calorically perfect gas case [3]. The measured drag coefficients for spheres covering subsonic, supersonic and hypersonic regimes are given in [53].

The flow past a sphere with a diameter $D = 12.7$ mm in a hypersonic air flow with equilibrium chemical reactions is considered. The flow parameters correspond to those used in experiments [50]. The inlet flow pressure is $p = 666.61$ Pa, and the inlet temperature is $T = 293$ K (the density is $\rho = 7.9 \times 10^{-3}$ kg/m^3). The Mach number varies in the range M $= 7.1–17.77$, which corresponds to the flight speed $V = 2438.4–6705.6$ m/s.

The computational domain is shown in the Figure 8. The distance be-

Table 2. Time (in seconds per 1000 time steps) and speedup for shock tube problem

No	Mesh 1			Mesh 2		
	CPU	GPU	Speedup	CPU	GPU	Speedup
A1	18.79	1.36	13.83	198.96	10.98	18.12
A2	27.79	1.60	17.35	261.63	13.31	19.65
B1	36.83	2.63	13.98	388.74	21.25	18.29
B2	56.28	3.14	17.90	515.97	26.28	19.63
No	Mesh 3			Mesh 4		
	CPU	GPU	Speedup	CPU	GPU	Speedup
A1	1918.32	99.60	19.26	17485.10	874.99	19.98
A2	2285.81	120.88	18.91	20542.40	914.83	22.46
B1	3776.11	197.58	19.11	35257.70	1736.47	20.30
B2	4672.10	240.95	19.39	40595.40	1822.97	22.27

tween outer boundary of the computational domain and stagnation point is 4 mm, and the distance between outer boundary of the computational domain and top sphere point is 8.65 mm. The problem is solved on the mesh containing 1.6×10^5 cells. The supersonic inlet boundary conditions are specified on the inlet boundary, and supersonic outflow boundary conditions are specified on the outlet boundary. Slip and adiabatic boundary conditions are applied to the sphere surface.

When the freestream Mach number is sufficiently high, the non-dimensional flow quantities become essentially independent of Mach number. Some typical results are shown in the Figure 9 for the fixed Mach number. Comparison of the results computed from the equilibrium chemically reacting case with similar results obtained from a calorically perfect gas with $\gamma = 1.4$ given in [3] for flow over a sphere show that the chemically reacting case exhibits higher densities and a thinner shock layer. Pressure reaches a maximum at the stagnation point and then decreases downstream. The high temperature effects on the flow variables become more pronounced with increasing free stream velocity, although the pressure is the least affected. The most intense dissociation occurs near the stagnation region, but the primary concentrations of nitrogen monoxide occur

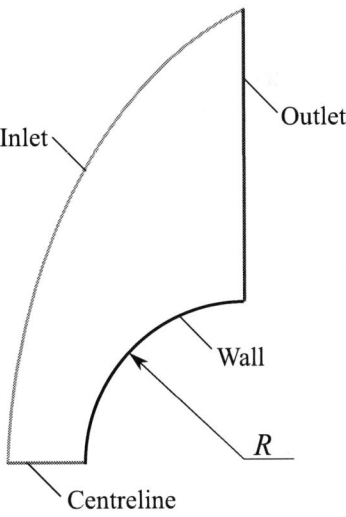

Figure 8. Computational domain and boundary conditions.

further downstream beyond the sonic line. Near the shock the mass fraction of oxygen is 0.22 and due to the high temperatures it dissociates and reacts with other air molecules as the flow moves downstream. Near the wall oxygen is almost completely dissociated but there is not a large amount of atomic oxygen.

The distributions of pressure, temperature and velocity along the stagnation line are shown in the Figure 10 (physical quantities are normalized on the free stream quantities). The lines 1 correspond to the model of perfect gas, and lines 2 correspond to the calculations with the model proposed in [47]. The ideal gas model overpredicts temperature in the shock layer because the internal energy is stored in the gas translational mode. The ideal gas model is a calorically perfect gas, and none of the other internal energy modes of the molecule are accounted for. Furthermore, the ideal N_2 remains as diatomic nitrogen and has not been allowed to dissociate which would otherwise absorb some of the internal energy as chemical energy. The temperature and non-dimensional pressure profiles along the stagnation line agree well with the axisymmetric solution presented in [55]. A small discrepancy is a consequence of the different treatment of physical quantities of equilibrium flow and the introduction of a more advanced

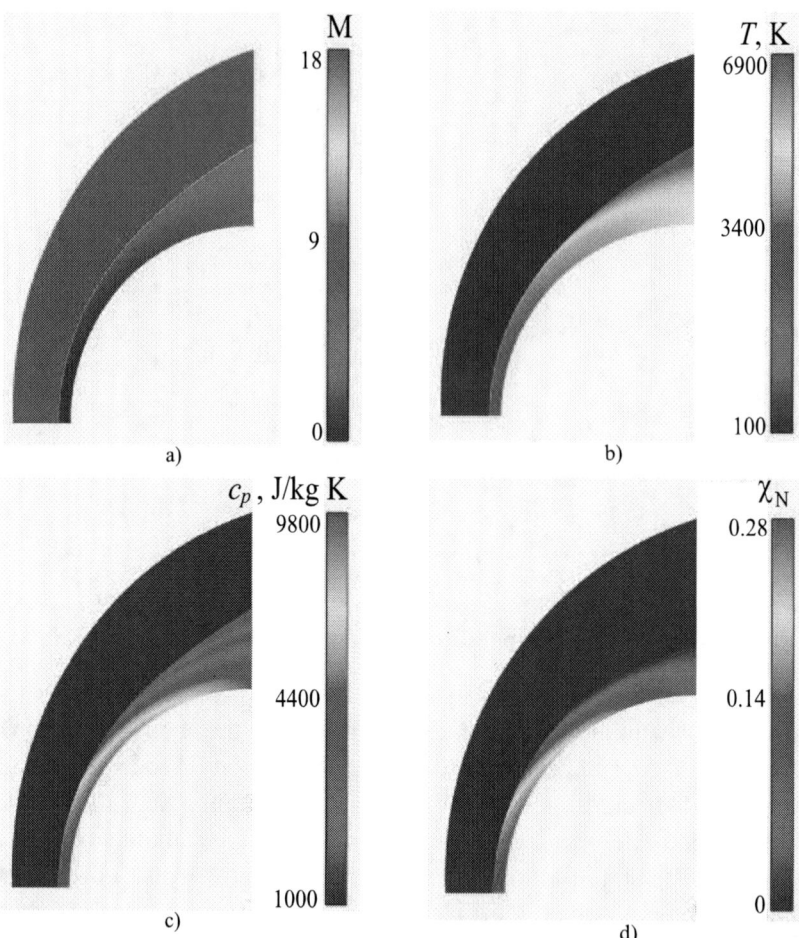

Figure 9. Contour plots of Mach number (a), temperatures (b), specific heat capacity at constant pressure (c) and mole fraction of atomic nitrogen (d) for $M = 17.77$.

upwind numerical scheme.

Pressure distributions on sphere surface show Figure 11. The gas pressure behind the shock undergoes almost a homogeneous variation in the process of the increasing Mach number. But the increasing of temperature is slow initially

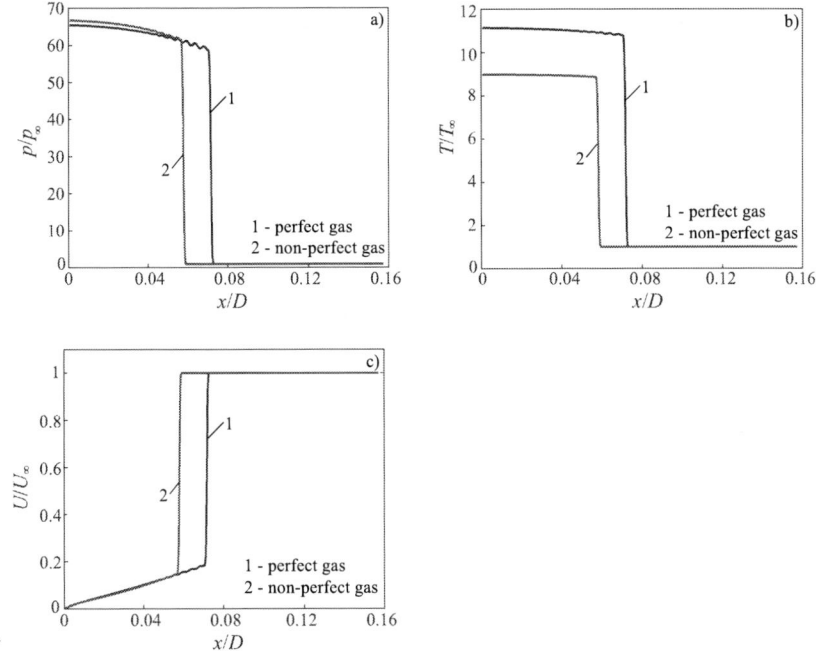

Figure 10. Distributions of pressure (a), temperature (b) and velocity (c) along the stagnation line for M = 7.106 (V = 2438.4 m/s).

and then becomes larger. The results are in a good agreement with those presented in [54].

Two important characteristics of the chemical equilibrium and perfect gas model distributions at the stagnation plane are concerned with the behind the shock flow variable magnitudes and the shock layer thickness. In comparison with the perfect gas model results, the results behind the shock wave obtained with the chemical equilibrium gas model are only slightly higher for pressure, considerably lower for temperature and higher for density. The higher density for the chemical equilibrium gas model leads to reduction of the shock standoff distance (shock layer thickness) in front of the sphere. The temperature, pressure and density behind the shock wave increase toward the sphere surface due to the continuous decrease in the flow velocity.

The density ratio has an important effect on the shock standoff distance in

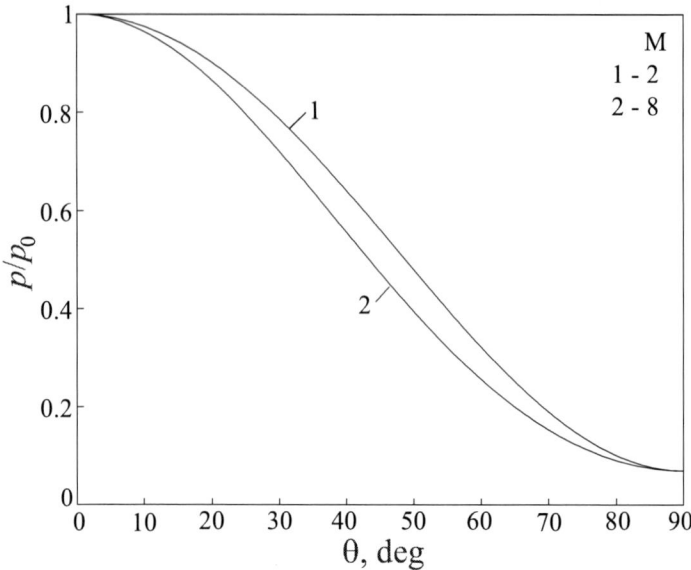

Figure 11. The ratio of the pressure at the sphere surface and the pressure at the stagnation point behind a normal shock wave for M = 2 (line 1) and M = 8 (line 2).

front of a sphere. An approximate expression for the shock standoff distance Δ on a sphere of diameter D is given by [50, 56]

$$\frac{\Delta}{D} = K\varepsilon, \quad \varepsilon = \frac{\gamma - 1}{\gamma + 1}\left(1 + \frac{2}{\gamma - 1}\frac{1}{M_\infty^2}\right),$$

where ε is the density ratio across the shock on the stagnation streamline, and K is the correction factor ($K = 0.39$–0.41). For high velocities, density ratio becomes small compared to unity. Therefore, the density ratio across a normal shock wave has a major impact on shock standoff distance. If the density ratio increases, then the shock standoff distance decreases. The chemical reactions increase the density ratio, which decreases the shock standoff distance. Therefore, in comparison to the calorically perfect gas results, the shock wave for a chemically reacting gas lies closer to the sphere (at the same velocity and altitude conditions).

Figure 12 shows that the shock standoff distance reduces with increasing free stream Mach number (increasing density behind the shock wave). Line 1 shows the shock standoff distance for air in full chemical equilibrium. Line 2 corresponds to the experimental measurements [50]. Line 3 gives the shock standoff distance for a perfect gas with $\gamma = 1.4$. The standoff distance decreases at first with the increasing Mach numbers, but it increases when Mach number becomes larger. The shock wave is closer to the wall for high Mach number. This is due to the fact that the temperature behind a shock wave increases with the Mach number increasing and goes up to induce the occurrence of chemical reactions for air components. Simultaneously, the pressure is also rising. However, the increase rates of the pressure and the temperature are not equal [3]. Since the calculated perfect gas normalized density values tend towards a finite maximum value of about 6 for air as the free stream Mach number increases, the shock wave thickness tends towards a minimum. For the Mach number about 10, the shock layer thickness is already close to this minimum since the normalized density is about 5.7. For the same free stream conditions, the chemical equilibrium gas shock layer is thinner than that for perfect gas and is continually changing with increasing free stream Mach number.

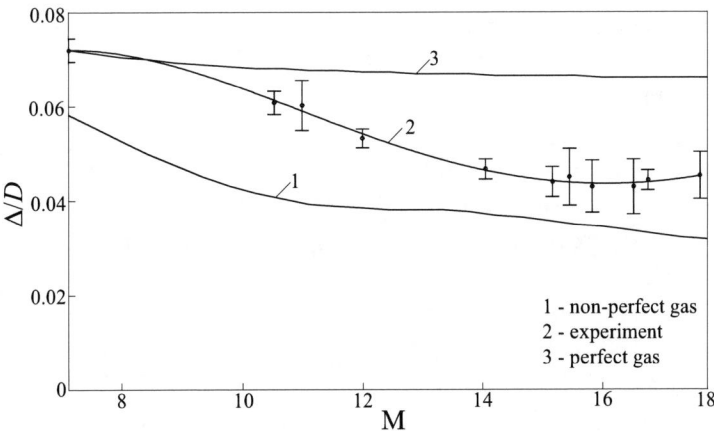

Figure 12. Dependence of the shock standoff distance on the free stream Mach number.

The shock standoff distance is directly proportional to the diameter of the sphere and inversely proportional to the density in the shock layer. For a perfect gas with $\gamma = 1.4$, the density ratio across the shock wave has a minimum value of 1/6. However, high velocity shock waves in air do not have such a limit, except at very low pressures where the relaxation times are large compared to the transit time for the air flowing over the sphere. The tendency is for the enthalpy to go into heat of formation in chemical reactions such as dissociation, rather than increasing the temperature of the air. The effect of dissociation is to increase the density and to decrease the shock detachment distance as compared to the perfect air result.

The shock standoff distance is smallest in the case of calorically perfect gas and the chemical non-equilibrium flow shock stand-off distance lies between calorically perfect gas and thermochemical equilibrium flow. In the case of calorically perfect gas, the kinetic energy of the flow ahead of the shock is mostly converted to translational and rotational molecular energy behind the shock. On the other hand, for a gas in thermal equilibrium or chemically reacting gas, the kinetic energy of the flow, when converted across the shock wave, is shared across all molecular modes of energy, and goes into the dissociation energy of the products of the chemical reactions. Hence, the temperature which is a measure of translational energy only, is less for such a case. In contrast, the pressure ratio is affected only by a small amount since pressure is a mechanically oriented variable and it is governed mainly by the fluid mechanics of the flow and not so much by the thermodynamics. The combined effect of pressure and temperature yields the density ratio across the shock which is more pronounced in case of a gas with internal energy excitation or chemically reacting gas. The shock standoff distance, which depends on the density ratio across the shock, therefore in this case is less as for non-reacting gas.

The computed drag coefficients for spheres are plotted in the Figure 13 against Mach number. Line 1 corresponds to the model of high-temperature air, and line 2 corresponds to the case of a perfect gas. Bullets shows measured drag coefficient from the ballistic range measurements [53]. The large drag rises in the subsonic regime associated with the drag-divergence phenomena and the decrease in drag in the supersonic regime. Drag approaches a plateau and becomes relatively independent of Mach number as Mach number becomes large. For a range of Mach numbers from 4 to 10, the drag coefficient is relatively constant at a value of 0.92.

Table 3 shows the execution time for various thermodynamic air models.

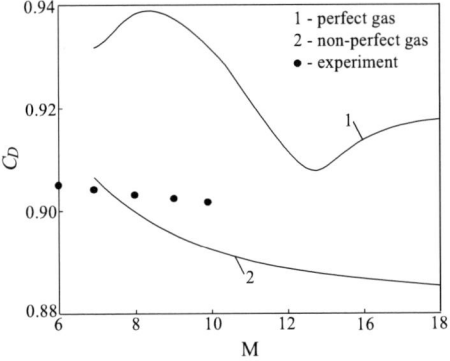

Figure 13. Dependence of the drag coefficient on Mach number.

Time discretization is based on the third-order Runge–Kutta scheme. Flux on the faces of the control volume are calculated with Godunov and Rusanov schemes [48]. A perfect gas model and Kraiko model proposed in [47] are used as thermodynamic model of air.

Table 3. Computational time and speedup for hypersonic sphere problem. Different thermodynamic models and computational options

Air model	Solver type	Server 1			Server 2		
		CPU	GPU	Speedup	CPU	GPU	Speedup
Perfect gas	Godunov	2.32	0.117	19.86	4.05	0.029	139.66
	Rusanov	1.78	0.106	16.79	2.85	0.027	105.41
Kraiko model	Godunov	45.96	0.242	189.91	84.08	0.061	1378.33
	Rusanov	45.42	0.232	195.76	81.94	0.058	1412.78

6.3. Hypersonic Aircraft Problem

In many hypersonic vehicles, the propulsion mechanism is integrated with the airframe [3]. In the integrated airframe propulsion concept for a hypersonic vehicle, the entire under surface of the vehicle is a part of the scramjet engine. Initial compression of the air takes place in the bow shock from the nose of the vehicle. Further compression and supersonic combustion occur inside a series of modules near the rear of the vehicle. The expansion of the combustion products is partially realized through nozzles in the engine modules and mainly over the bottom rear surface of the vehicle. The lift is produced by high pressure behind the bow shock wave and exerted on the relatively flat under surface of the vehicle. Integration of airframe and propulsion system makes the modelling and control of hypersonic vehicles challenging. As temperature rises because of compression and viscous effects, translational and vibrational motion of the fluid molecules are high enough to cause dissociation. Monitoring the dissociation of molecular oxygen molecules is most important as scramjets use molecular oxygen for the combustion oxidizer. If dissociation occurs, combustion process is limited in terms of available energy for the production of thrust. Hypersonic inlets must be appropriately modelled to determine chemical composition and the physical state of the air entering the combustion chamber.

The flow past of a hypersonic aircraft at zero angle of attack is considered. The model, shown in the Figure 14, approximately corresponds to NASA hypersonic aircraft X-43 presented in [4, 5, 9]. Full geometrical dimensions are unknown, and geometrical model is designed using some dimensions given in [4, 57]. The upper body of the vehicle is simply a flat surface, which is kept at zero angle of attack for simplicity. The lower surface consists of a frontal wedged surface, a scramjet engine with a constant cross sectional area and another trailing wedged surface. The frontal wedged surface serves as a diffuser for the flow entering the scramjet, and the trailing surface acts as a propulsive surface. The aircraft three views and main dimensions are shown in the Figure 15 (dimensions are given in millimeters). The leading edge angle is $5°$, the length of the engine 0.68 m, and the engine cross-section area is 0.3 m (height) by 0.5 m (span). The frontal area of the model is about 0.153 m^2. The flight conditions correspond to the Mach number $M = 10$ at an altitude of 30 km (the flight speed is $V = 2976.62$ m/s). At the given altitude, the atmospheric pressure and temperature are $p = 1172$ Pa and $T = 227$ K, respectively (density is 0.0184 kg/m^3 and dynamic viscosity is 0.148×10^{-4} Pa·s). The results of wind tunnel tests are published in [57].

The problem is solved in a three-dimensional formulation for half of the computational domain whose dimensions are chosen equal to $6 \times 4 \times 2$ m^3. At the inlet boundary, the supersonic inflow conditions are specified, and supersonic outflow conditions are applied to outlet boundary. Slip and no-penetration boundary conditions are used on the walls. The surface of the body is assumed

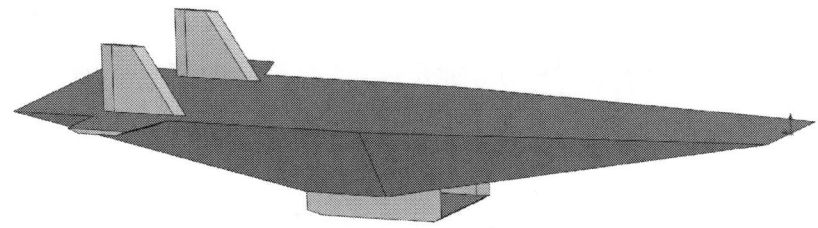

Figure 14. Model of hypersonic aircraft.

to be adiabatic.

The unstructured computational mesh is constructed using the Numeca HEXPRESS package, which allows obtaining a qualitative partially orthogonal grid with a controlled thickening of nodes in the regions of large gradients of physical quantities. The mesh consists of 16 million cells and 48 million faces. The mesh contains 439221 boundary faces.

The distributions of the velocity magnitude, temperature, molar concentration of nitrogen oxide and effective heat capacity ratio in the plane of symmetry are shown in the Figure 16. Due to the fact that the aircraft has well streamlined shape, the gas temperature does not increase too much and relatively little nitrogen oxide is formed, while other components are practically not formed. Mole fractions of molecular oxygen and nitrogen oxide decrease as temperature past the shock wave increases. Therefore, the mole fraction of atomic nitrogen increases.

The temperature distribution on the aircraft surface is shown in the Figure 17. Level lines of the density gradient are shown in the cutting planes. The highest temperature is observed at the inlet to the air intake. The oblique shock wave created at the nose of aircraft propagates downstream underneath the vehicle and can impinge on the leading edge of the cowl of the scramjet engine. The inlet cowl has a blunt leading edge to reduce aerodynamic heating, and therefore a detached curved shock wave exists just upstream of the cowl leading edge similar to the blunt body flows. The shock from the nose of the vehicle does not impinge directly on the surface of the cowl, but rather on the cowl shock wave, setting up a rather complex shock-shock interaction that can change the nature of both shocks and the surrounding flow field.

Pressure distributions and limiting streamlines on the aircraft surface are

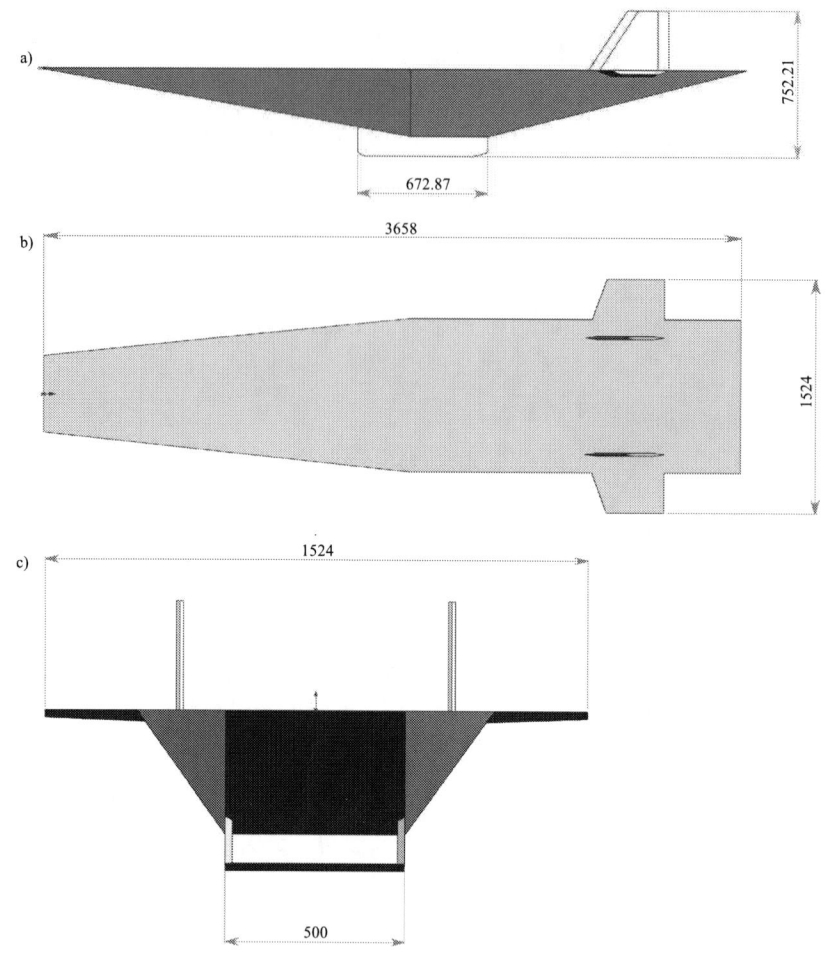

Figure 15. Three views of the hypersonic vehicle.

presented in the Figure 18. Limiting streamlines show location of vortex structures

Drag and lift coefficients as functions of angle of attack are shown in the Figure 19. The angle of attack varies from −10 to +10 degrees. In the range from 0 to 10 degrees, increase in angle of attack leads to increase in aerodynamic quality (ratio of lift and drag). Quantitative comparison with experimental data

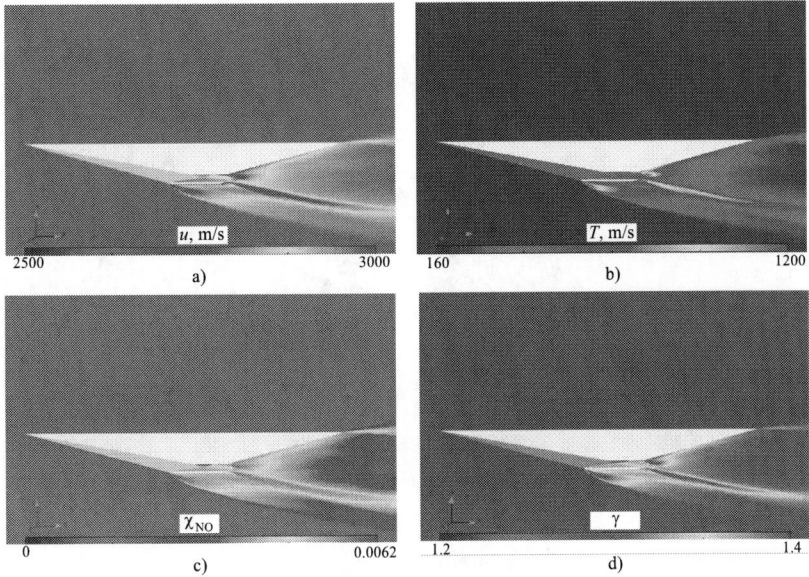

Figure 16. Contours of velocity magnitude (a), temperature (b), molar fraction of oxide nitrogen (c) and effective ratio of specific heat capacities (d) in symmetry plane.

is difficult due to different shapes of hypersonic aircrafts, different flow regimes and reference parameters. However, qualitative dependances of drag and lift coefficients on angle of attack are similar to those observed experimentally. Dependance of drag coefficient is close to parabolic dependance, and dependance of lift coefficient is close to linear dependance.

To evaluate performance of the computational procedure, the model of perfect air without chemical reactions is implemented. Table 4 shows the execution time of one step with averaging over 10 iterations for various thermodynamic air models. Time discretization is based on the third-order Runge–Kutta scheme. Flux on the faces of the control volume are calculated with Godunov scheme [48]. A perfect gas model and Kraiko model proposed in [47] are used as thermodynamic model of air. CPU code is not organized optimally with the Godunov scheme. Using GPU allows to get a significant acceleration compared to a single CPU core. The use of GPU leads to 10 times faster calculations compared to the calculations on CPU.

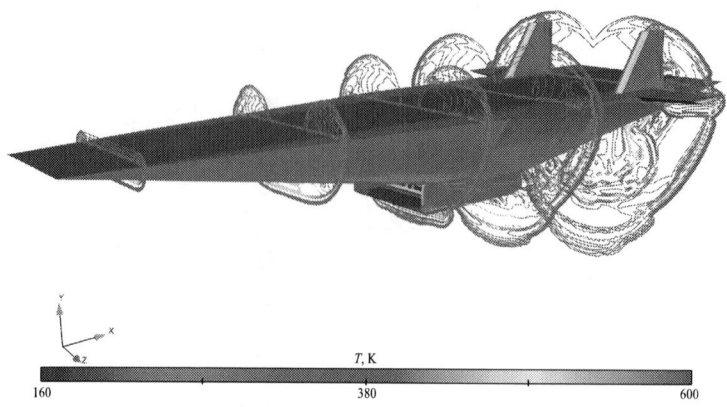

Figure 17. Contours of temperature and density gradient.

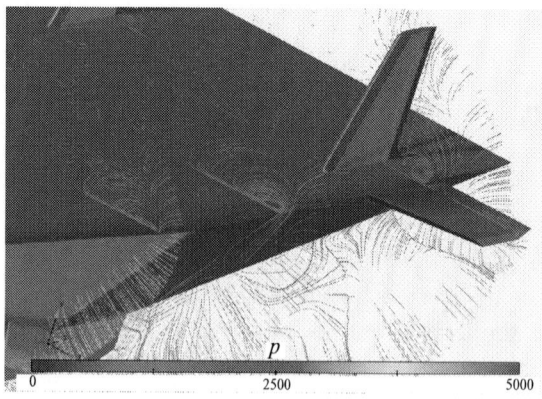

Figure 18. Contours of pressure and limiting streamlines.

CONCLUSION

The finite volume method was applied to solve full Navier–Stokes equations on unstructured meshes, and CUDA technology was used for programming implementation of parallel computational algorithms. The developed CFD code

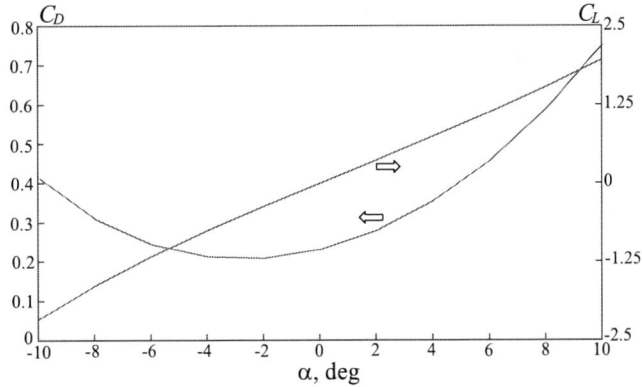

Figure 19. Drag and lift as functions of angle of attack for M = 6.

Table 4. Computational time and speedup for hypersonic aircraft problem. Different thermodynamic models

Air model	Server 1			Server 2		
	CPU	GPU	Speedup	CPU	GPU	Speedup
Perfect gas	134.40	8.61	15.61	212.42	1.80	117.81
Kraiko model	5009.49	19.76	253.47	3695.56	4.01	922.74

is able to simulate hypersonic flows with a reasonable degree of confidence. The numerical results show that the high-temperature gas effects significantly change flowfield including the standoff distance of bow shock over the front part of sphere and other flow properties. The ability of the computations to accurately capture the shock shapes and the standoff distances demonstrates the capability of the code to model flows with high-temperature gas effects.

GPUs have evolved as a new paradigm for scientific computations. Cost/performance ratio and low power consumption make GPUs attractive for high-resolution CFD computations. However, in order to exploit the inherent architecture of the device, the numerical algorithm as well as data structures are carefully tailored to minimize the memory access and any recursive relations in the computational algorithm. Speedup of solution on GPU with respect to solution on CPU is compared with the use of various meshes and computational options. Performance measurements show that numerical schemes developed achieve 20 to 50 speedup on GPU hardware compared to CPU reference implementation. The results obtained provide promising perspective for designing a GPU-based software framework for applications in CFD.

ACKNOWLEDGMENT

The study was financially supported by the Russian Science Foundation (project No. 19-71-10019).

REFERENCES

[1] Smirnov, N.N. (2017). Ensuring safety of space flights. *Acta Astronautica*, 135, 1–5.

[2] Silnikov, M.V., Guk, I.V., Nechunaev, A.F. and Smirnov, N.N. (2018). Numerical simulation of hypervelocity impact problem for spacecraft shielding elements. *Acta Astronautica*, 150, 56–62.

[3] Anderson, J.D. (2006). Hypersonic and high temperature gas dynamics. *AIAA*, 811 p.

[4] Huebner, L.D., Rock, K.E., Ruf, E.G., Witte, D.W. and Andrews, E.H. (2001). Hyper-X flight engine ground testing for flight risk reduction. *Journal of Spacecraft and Rockets*, 38(6), 844–852.

[5] Reubush, D.E., Nguyen, L.T. and Rausch, V.L. (2003). Review of X-43A return to flight activities and current status. *AIAA Paper*, 2003-7085.

[6] Borovoi, V.Ya., Skuratov, A.S. and Surzhikov, S.T. (2004). Study of convective heating of segmental-conical martian descent vehicle in shock wind tunnel. *AIAA Paper*, 2004-2634.

[7] Deiwert, G., Strawa, A., Sharma, S. and Park, C. (1989). *Experimental program for real gas flow code validation at NASA Ames Research Centre.* Washington, NASA.

[8] Wang, Z., Sun, X., Huang, W., Li, S. and Yan, L. (2016). Experimental investigation on drag and heat flux reduction in supersonic/hypersonic flows: a survey. *Acta Astronautica*, 129, 95–110.

[9] Mirmirani, M., Wu, C., Clark, A., Choi, S. and Fidan, B. (2005). Air-breathing hypersonic flight vehicle modeling and control, review, challenges, and a CFD-based example. *Proceeding of the Workshop on Modeling and Control of Complex Systems*, Ayia Napa, Cyprus, 30 June — 1 July.

[10] Yoon, S., Gnoffo, P.A., White, J.A. and Thomas, J.L. (2007). Computational challenges in hypersonic flow simulations. *AIAA Paper*, 2007-4265.

[11] Schmisseur, J.D. (2015). Hypersonics into the 21st century: a perspective on AFOSR-sponsored research in aerothermodynamics. *Progress in Aerospace Sciences*, 72, 3–16.

[12] Stalker, R. (1989). Hypervelocity aerodynamics with chemical nonequilibrium. *Annual Review of Fluid Mechanics*, 21, 37–60.

[13] Tirsky, G. (1993). Up-to-date gasdynamic models of hypersonic aerodynamics and heat transfer with real gas properties. *Annual Review of Fluid Mechanics*, 25, 151–181.

[14] Hilsenrath, J. and Klein, M. (1965). *Tables of thermodynamic properties of air in chemical equilibrium including second virial corrections from 1500 to 15,000 K*. Tullahoma, Arnold Engineering Development Center, AEDC-TR-65-68.

[15] Bertin, J. and Cummings, R. (2006). Critical hypersonic aerothermodynamic phenomena. *Annual Review Fluid Mechanics*, 38, 129–157.

[16] Coblish, J., Smith, M., Hand, T., Candler, G. and Nompelis, I. (2005). Double-cone experiment and numerical analysis at AEDC Hyper-velocity wind tunnel No 9. *AIAA Paper*, 2005-0902.

[17] Tissera, S., Titarev, V. and Drikakis, D. (2010). Chemically reacting flows around a double cone, including ablation effects. *AIAA Paper*, 2010-1285.

[18] Chanetz, B. (2012). Low and high enthalpy shock-wave/boundary layer interactions around cylinder-flare models. *Progress in Flight Physics*, 3, 107–118.

[19] Candler, G.V. and Nompelis, I. (2002). CFD validation for hypersonic flight: real gas flows. *NATO Report*, RTO-TR-AVT-007-V3.

[20] MacCormack, R.W. (2013). Carbuncle computational fluid dynamics problem for blunt-body flows. *Journal of Aerospace Information Systems*, 10(5), 229–239.

[21] Menne, S., Weiland, C., D'Ambrosio, D. and Pandolfi, M. (1995). Comparison of real gas simulations using different numerical methods. *Computers and Fluids*, 24(3), 189–208.

[22] Boulahia, A., Abboudi, S. and Belkhiri, M. (2014). Simulation of viscous and reactive hypersonic flows behaviour in a shock tube facility: TVD schemes and flux limiters application. *Journal of Applied Fluid Mechanics*, 7(2), 315–328.

[23] Lockwood, B. and Mavriplis, D. (2013). Gradient-based methods for uncertainty quantification in hypersonic flows. *Computers and Fluids*, 85, 27–38.

[24] Tchuen, G., Burtschell, Y. and Zeitoun, D.E. (2008). Computation of non-equilibrium hypersonic flow with artificially upstream flux vector splitting (AUFS) schemes. *International Journal of Computational Fluid Dynamics*, 22(4), 1061–8562.

[25] Gimelshein, S., Markelov, G. and Ivanov, M. Real gas effects on the transition between regular and Mach reflections in steady flows. *AIAA Paper*, 1998-877.

[26] Park, C. (1990). *Nonequilibrium hypersonic aerothermodynamics*. John Wiley & Sons.

[27] Nompelis, I., Candler, G.V., Holden, M.S. and Wadhams, T.P. (2003). Computational investigation of hypersonic viscous-inviscid interactions in high-enthalpy flows. *AIAA Paper*, 2003-3642.

[28] Youssefi, M.R. and Knight, D. (2017). Assessment of CFD capability for hypersonic shock wave laminar boundary layer interactions. *Aerospace*, 4(2), 1–18.

[29] Holden, M.S. (1998). Shock interaction phenomena in hypersonic flows. *AIAA Paper*, 98-2751.

[30] Ma, Y. and Zhong, X. (1999). Numerical simulation of transient hypersonic flow with real gas effects. *AIAA Paper*, 99-16309.

[31] Huang, W., Yan, L., Liu, J. and Tan, J. (2015). Drag and heat reduction mechanism in the combinational opposing jet and acoustic cavity concept for hypersonic vehicles. *Aerospace Science and Technology*, 42, 407–414.

[32] Park, C. and Yoon, S. (1991). Fully-coupled implicit method for thermochemical nonequilibrium air at sub-orbital flight speeds. *Journal of Spacecraft and Rockets*, 28(1), 31–39.

[33] Chen, S., Hu, Y. and Sun, Q. (2012). Study of the coupling between real gas effects and rarefied effects on hypersonic aerodynamics. *AIP Conference Proceedings*, 1501, 1515–1521.

[34] Chinnappan, A.K., Malaikannan, G. and Kumar, G. (2017). Insights into flow and heat transfer aspects of hypersonic rarefied flow over a blunt body with aerospike using direct simulation Monte-Carlo approach. *Aerospace Science and Technology*, 66, 119–128.

[35] Owens, J.D., Luebke, D., Govindaraju, N., Harris, M., Krűger, J., Lefohn, A.E. and Purcell, T.J. (2007). A survey of general-purpose computation on graphics hardware. *Computer Graphics Forum*, 26(1), 80–113.

[36] Sanders J., Kandrot E. (2011). *CUDA by example: an introduction to general-purpose GPU programming.* Boston: Pearson Education.

[37] Fu, L., Gao, Z., Xu, K. and Xu, F. (2014). A multi-block viscous flow solver based on GPU parallel methodology. *Computers and Fluids*, 95, 19–39.

[38] Jacobsen, D.A. and Senocak, I. (2013). Multi-level parallelism for incompressible flow computations on GPU clusters. *Parallel Computing*, 39(1), 1–20.

[39] Tuttafesta, M., Colonna, G. and Pascazio, G. (2013). Computing unsteady compressible flows using Roe's flux-difference splitting scheme on GPUs. *Computer Physics Communications*, 184(6), 1497–1510.

[40] Kampolis, I.C., Trompoukis, X.S., Asouti, V.G. and Giannakoglou, K.C. (2010). CFD-based analysis and two-level aerodynamic optimization on

graphics processing units. *Computer Methods in Applied Mechanics and Engineering*, 199(9–12), 712–722.

[41] Emelyanov, V.N., Karpenko, A.G., Kozelkov, A.S., Teterina, I.V., Volkov, K.N. and Yalozo, A.V. (2017). Analysis of impact of general-purpose graphics processor units in supersonic flow modelling. *Acta Astronautica*, 135, 198–207.

[42] Emelyanov, V., Karpenko, A. and Volkov, K. (2017). Development and acceleration of unstructured mesh based CFD solver. *Progress in Flight Physics*, 9, 223–244.

[43] Rathakrishnan, E. (2015). *High enthalpy gas dynamics*. Wiley Online Library.

[44] Gupta, R. N., Scott, C. D. and Moss, J. N. (1985). Slip-boundary equations for multicomponent nonequilibrium airflow. *NASA Technical Report*, TP-1985-2452.

[45] Lofthouse, A.J., Scalabrin, L.C. and Boyd, I.D. (2008). Velocity slip and temperature jump in hypersonic aerothermodynamics. *Journal of Thermophysics and Heat Transfer*, 22(1), 38–49.

[46] Chase, M.W., Curnutt, J.L., McDonald, R.A. and Syverud, A.N. (1978). JANAF thermochemical tables. *Journal of Physical and Chemical Reference Data*, 7(3), 793–940.

[47] Kraiko, A.N. and Makarov, V.E. (1996). Explicit analytic formulas defining the equilibrium composition and thermodynamic functions of air for temperatures from 200 to 20000 K. *High Temperature*, 34(2), 202–213.

[48] Toro, E.F. (2009). Riemann solvers and numerical methods for fluid dynamics: a practical introduction. Springer.

[49] Roe, P.L. (1997). Approximate Riemann solvers, parameter vectors, and difference schemes. *Journal of Computational Physics*, 135(2), 250–258.

[50] Lobb, R.K. (1964). *Experimental measurement of shock detachment distance on spheres fired in air at hypervelocities / High Temperature Aspects of Hypersonic Flow*. Oxford, Pergamon Press, 519–527.

[51] McIntyre, T.J., Bishop, A.I., Rubinsztein-Dunlop, H. and Gnoffo, P. (2003). Comparison of experimental and numerical studies of ionizing flow over a cylinder. *AIAA Journal*, 41(11), 2157–2161.

[52] Olivier, H. (2000). A Theoretical model for the shock stand-off distance in frozen and equilibrium flows. *Journal of Fluid Mechanics*, 413, 345–353.

[53] Cox, R.N. and Crabtree, L.F. (1965). *Elements of hypersonic aerodynamics*. New York, Academic Press.

[54] Holt, M. and Hoffman, G.H. (1961). Calculation of hypersonic flow past sphere and ellipsoids. *American Rocket Society*, 61-209-1903.

[55] Olsen, M.E., Liu, Y., Vinokur, M. and Olsen, T. (2004). Implementation of premixed equilibrium chemistry capability in OVERFLOW. *AIAA Paper*, 2004-1273.

[56] Lomax, H. and Inouye, M. (1964). Numerical analysis of flow properties about blunt bodies moving at supersonic speeds in an equilibrium gas. *NASA Technical Report*, TR-R-204.

[57] Engelund, W., Holland, S., Cockrell, C. and Bittner, R. (1999). Propulsion system airframe integration issues and aerodynamic database development for the Hyper-X flight research vehicle. *Proceedings of the 14th International Symposium on Air-breathing Engines*, Florence, Italy, 5–10 September 1999, ISABE 99-7215.

INDEX

A

ablation, 2, 4, 5, 9, 18, 20, 21, 22, 25, 27, 28, 30, 31, 33, 34, 36, 50, 52, 53, 165
ablation phase change, 18
ablative, viii, 2, 3, 4, 5, 7, 8, 9, 18, 21, 22, 27, 32, 33, 34, 40
ablative phase change, viii, 2, 3, 4, 32

B

blast furnace(s), viii, 55, 56, 57, 79, 85
burner(s), viii, 55, 56, 63, 64, 66, 68, 69, 70, 73, 75, 76, 77, 78, 79, 80, 81, 82, 85, 86

C

carbon based ablative materials, 5
carbon dioxide emissions, 57, 101
carbon monoxide, 58, 71
CFD simulation, 11, 56, 79, 129
changed phase, 46
chemical kinetic, 8
CO_2 capture, 98
cokemaking, 57

compressible flow, 2, 40, 50, 52, 167
computational fluid dynamics (CFD), v, vii, viii, ix, 1, 2, 3, 9, 11, 17, 23, 52, 54, 55, 56, 61, 66, 67, 79, 85, 87, 88, 91, 92, 93, 94, 96, 97, 98, 125, 126, 127, 128, 129, 130, 136, 140, 143, 146, 148, 162, 163, 164, 165, 166, 167, 168
concentrate, viii, 55, 56, 57, 59, 60, 62, 63, 64, 68, 71, 72, 73, 76, 77, 88, 92, 96
condensation, 2, 10, 11, 13, 14, 15, 23, 24, 26, 27, 28, 42, 51, 53, 54, 96, 101
contact angle, 23, 27, 100

D

design, v, vii, viii, ix, 55, 56, 67, 68, 73, 74, 75, 76, 77, 79, 82, 84, 85, 88, 90, 125, 126, 127, 129
direct reduction processes, 56
direct steelmaking, 57, 58, 67, 79
DRI, 79
droplet deformation, 11, 26, 40, 43
droplets, vii, viii, 2, 4, 15, 18, 23, 26, 40, 41, 42, 43, 50, 54

E

energy consumption, viii, 55, 56, 57, 101
energy loss, 73
energy saving, 56
energy source, 13, 16, 25, 29, 128
enthalpy-porosity, 15, 17, 29
enthalpy-porosity method, 15
Eulerian frame, 71
evaporated, 25, 41
evaporation, viii, 2, 4, 9, 10, 11, 12, 14, 15, 22, 23, 24, 25, 26, 27, 28, 29, 31, 40, 42, 43, 44, 45, 51
evaporation/condensation, 4, 9, 11, 12, 22, 23, 26, 27, 28, 29, 51
evaporative, 3, 10, 22, 32, 43
evaporative phase, 22, 32
evaporative/condensation phase change, 3

F

flash ironmaking, vii, viii, 55, 56, 57, 59, 61, 62, 67, 68, 69, 75, 79, 80, 85, 91, 94
flash ironmaking technology (FIT), viii, 55, 56, 57, 58, 59, 67, 79, 85, 89, 90, 91, 92, 94
flash reactor, viii, 55, 56, 60, 64, 74, 85
flash reduction, 56, 61, 67
flow field, viii, 3, 31, 53, 56, 73, 85, 128, 159
fluidized-bed processes, 56

G

gas generation, 22, 33, 40
gas species production rate, 9
graphics, ix, 125, 126, 129, 167, 168
greenhouse-gas emissions, 56

H

heat loss, viii, 56, 73, 74, 78, 79, 82, 84, 85, 86
heat of vaporization, 11, 23
hematite concentrate, 58, 60
hollow fiber membrane, vii, viii, 97, 98, 102
hydrogen, 26, 40, 56, 57, 58, 60, 61, 62, 71, 80, 87, 88, 92, 95, 96, 101, 122
hypersonic flow, x, 125, 126, 127, 128, 129, 130, 131, 135, 148, 163, 164, 165, 166, 167, 169

I

industrial, vii, viii, 1, 2, 55, 56, 60, 66, 79, 80, 81, 85, 101
industrial reactor(s), vii, viii, 55, 56, 66, 79, 80, 81, 85
interface area, 12, 22, 23
interface(s), 3, 10, 11, 12, 13, 15, 17, 18, 22, 23, 24, 25, 26, 28, 29, 31, 40, 41, 42, 46, 51, 63, 64, 100, 140, 148
iron oxide concentrate, 56

K

kinetic parameters, 59
kinetics, 56, 58, 85, 87, 88, 91, 92, 93, 95, 96, 138

L

Lagrangian mode, 71
laminar-flow reactor(s), 59, 60
latent heat, 2, 10, 13, 14, 16, 23, 27, 28, 29
latent heat content, 16
latent heat of fusion, 14, 16, 28
latent heat of vaporization, 10, 13, 23, 27, 28

Index

M

magnetite, 56, 58, 59, 60, 64, 66, 78, 80, 87, 88, 90, 91, 92, 93, 95, 96
magnetite concentrate, 58, 59, 60, 64, 66, 78, 80, 87, 88, 90, 91, 92, 93, 95, 96
mass flux, 6, 8, 18, 20
mass flux rates, 6
mass transfer, ix, 12, 13, 14, 23, 98, 102, 103, 104, 110, 112, 113, 114, 117, 118, 120, 121
mass transfer rates, 12
melting, 2, 3, 4, 14, 15, 17, 27, 28, 29, 30, 31, 45, 47, 51, 54, 77, 84
melting and solidification, 14, 27, 28, 30, 45
melting/solidification, 3, 4, 14, 15, 27, 30, 47
melting/solidification phase change, 3, 14, 15
metallization, 71, 72, 79, 82, 85
methane, 61, 88, 91
modeling, v, vii, 2, 4, 5, 9, 11, 12, 14, 17, 18, 21, 22, 23, 25, 27, 29, 30, 31, 32, 33, 40, 42, 43, 45, 47, 50, 51, 52, 53, 54, 67, 87, 91, 93, 96, 97, 98, 123, 165
mono-ethanol amine (MEA), v, vii, ix, 97, 98, 99, 100, 102, 107, 108
mushy zone, 15, 16

N

natural gas, 56, 58, 61, 62, 65, 69, 71, 76, 80, 81, 83, 84, 101

O

operating conditions, 43, 66, 67, 71, 72, 73, 76, 80
oxidation and nitridation reactions, 5
oxy-fuel burner, 61

P

parallel algorithm, 126, 142
partial combustion, 56, 58, 61, 71, 74, 77
particle distribution, 79, 85
particles, vii, viii, 2, 4, 18, 21, 22, 25, 27, 30, 31, 32, 33, 34, 40, 45, 46, 47, 50, 51, 53, 56, 57, 58, 59, 60, 64, 68, 73, 76, 77, 78, 79, 83, 84, 85, 87, 88, 92, 93, 95, 96, 132, 133, 134
pelletization, 57
phase change, vii, 1, 2, 3, 4, 5, 9, 10, 11, 12, 13, 14, 15, 16, 17, 18, 22, 23, 24, 25, 27, 28, 29, 32, 40, 42, 47, 48, 49, 50, 51, 52, 53, 54
pilot flash reactor, 62, 63
processor unit, vii, 126, 144, 168
pyrolysis, 8, 22

R

rate equations, 59, 60
reactor design, 56, 67, 69, 73
regression, 5, 7, 9, 18, 28, 29, 30, 31, 50
regression rate, 7, 18

S

saturation temperature, 10, 12, 13, 23, 24
scale-up, 67
sensible enthalpy, 16
shaft furnace processes, 56
simulation, v, vii, ix, 17, 22, 32, 33, 34, 40, 43, 47, 52, 53, 54, 61, 69, 71, 76, 85, 88, 91, 94, 97, 98, 102, 103, 104, 125, 126, 127, 128, 129, 130, 131, 133, 135, 136, 137, 139, 141, 143, 145, 147, 148, 149, 151, 153, 155, 157, 159, 161, 163, 164, 165, 166, 167, 169
sintering, 57

Sohn, v, 55, 56, 58, 60, 86, 87, 88, 89, 90, 91, 92, 93, 94, 95, 96
solidification, viii, 2, 14, 15, 17, 27, 28, 30, 32, 45, 46, 47, 48, 49, 50, 51, 53
solidification and melting, 14, 15, 51, 53
species and rate of gas generation, 5
surface tension, 11, 12, 23, 27, 31, 41, 52, 100

T

thermal conditions, 2, 3, 9, 10, 23, 46, 47
thermal/transport properties, 14
thermodynamic properties, 14, 127, 165
third body efficiencies, 9
tonnage oxygen, 57, 62

U

unstructured mesh, 126, 129, 136, 137, 162, 168

V

vapor, 10, 11, 12, 13, 23, 24, 25, 26, 40, 41, 42, 43, 93, 95, 102
vaporization/evaporation, 11
volume fraction, 11, 12, 22, 100
volume of fluid, vii, viii, 2, 11, 54
volumetric energy source, 20, 25, 29
volumetric generation, 13, 35, 39
volumetric mass generation, 12, 13, 20, 24
volumetric sources, 16, 20